Introductory Crop Science

Introductory Crop Science

Lauren Ingram

Callisto Reference

www.callistoreference.com

Callisto Reference,
118-35 Queens Blvd., Suite 400,
Forest Hills, NY 11375, USA

Visit us on the World Wide Web at:
www.callistoreference.com

ISBN: 978-1-64116-526-6 (Hardback)

Cataloging-in-Publication Data

Introductory crop science / Lauren Ingram.
p. cm.
Includes bibliographical references and index.
ISBN 978-1-64116-526-6
1. Crop science. 2. Crops. 3. Agriculture. I. Ingram, Lauren.
SB91 .I58 2022
630--dc23

TABLE OF CONTENTS

PREFACE

This book aims to help a broader range of students by exploring a wide variety of significant topics related to this discipline. It will help students in achieving a higher level of understanding of the subject and excel in their respective fields. This book would not have been possible without the unwavered support of my senior professors who took out the time to provide me feedback and help me with the process. I would also like to thank my family for their patience and support.

The scientific knowledge, learning, practice and understanding related to the management, cultivation and production of crops is known as crop science. There are various types of crops which are dealt with in this field such as food crops, cash crops and horticulture crops. Crop science makes use of principles from a diverse range of disciplines such as breeding, genetics, management and production. The application of scientific and organized knowledge for solving practical problems is known as crop technology. Some of the technological interventions which fall within this discipline are seed technologies, agronomic technologies and hybrid technologies. This book unfolds the innovative aspects of crop science which will be crucial for the progress of this field in the future. The topics included herein on in this field are of utmost significance and bound to provide incredible insights to readers. Coherent flow of topics, student-friendly language and extensive use of examples make this book an invaluable source of knowledge.

A brief overview of the book contents is provided below:

Chapter - Crop and Crop Science

Any plant product that can be grown and harvested for profit and subsistence is known as a crop. The study of various kinds of crops and their environment is known as crop science. There are different types of crops which are studied within this field such as cash crops and catch crops. This chapter will briefly introduce the diverse aspects of these types of crops and crop science.

Chapter - Crop Genetics and Breeding

The study of genetic variation, genes and heredity in crops is known as crop genetics. Crop breeding is the branch of science that is concerned with altering the traits of plants in order to produce desired characteristics in crops. This chapter has been carefully written to provide an easy understanding of the different techniques and technologies used within crop genetics and crop breeding.

Chapter - Crop Production Systems

On the basis of the type of crop and use, the agricultural production systems are classified into various practices such as mixed farming, subsistence farming, plantation farming, shifting cultivation, intercropping, monocropping, companion planting, etc. This chapter discusses in detail these crop production systems.

Chapter - Crop Diseases

Crops are affected by numerous diseases such as botrytis cinerea, Phomopsis stem, pod blight, common scab, rust disease, Ascochyta blight and bacterial blight. A few of the soil based crop diseases are rhizoctonia, verticillium and crown rot. All these crop diseases have been carefully analyzed in this chapter.

Chapter - Crop Protection

The science and practice of managing plant diseases and pests that harm agricultural crops is referred to as crop protection. Some of the chemicals and biological compounds that are used to manage crop are fungicides, insecticides, biopesticides, etc. The chapter closely examines these chemicals used in crop protection to provide an extensive understanding of the subject.

Chapter - Crop Management

Crop management refers to the agricultural practices which are used to improve the development, growth and yield of agricultural crops. Some of the important practices associated with it are water management and crop rotation. All these diverse aspects of crop management have been carefully analyzed in this chapter.

Lauren Ingram

Crop and Crop Science

CHAPTER 1

Any plant product that can be grown and harvested for profit and subsistence is known as a crop. The study of various kinds of crops and their environment is known as crop science. There are different types of crops which are studied within this field such as cash crops and catch crops. This chapter will briefly introduce the diverse aspects of these types of crops and crop science.

CROP

A crop is an organism deliberately grown with proper care and management and harvested for obtaining yield. As for example, cultivated plants, poultry, cattle, fishes, shrimp, honey bee, silk worm, lac insect etc.

Crop plants refer to those plants which are intentionally grown by men with proper care with a view to fulfilling the requirement of certain needs like food, feed, cloth, oil, timber, shelter etc. Thus men are directly or indirectly benefited by growing crops.

Generally, agronomic crops are grown together collectively and harvested in bulk. They are also known as field crop.

Classification of Crop

Classification is an acceptable product of scientific studies and is obviously based on well-defined characteristics or concepts that separate one group from another.

The agrarian/agricultural classification of crops based upon utilization is given below:

- Cereal: Ceres (a Roman word) was the name of a Roman goddess who was the 'giver of grain'. Cereals are the cultivated grasses grown for their edible starchy grains (one seeded fruit known as caryopsis). In general, the larger grains used as staple food are cereals such as rice, wheat, maize, barley, sorghum. Millets are the small grained cereals which are of minor importance as food and they have a single cover (due to fusion of testa to the pericarp) such as bajri. Millets are also used as staple food in drier regions of the developing countries.
- Oil seed: Crop seeds that are rich in fatty acids, are used to extract vegetable oil

to meet various requirements, for instance mustard, rape, sesame, sunflower, safflower, castor, linseed, groundnut, soybean.

- Pulse: Seeds of leguminous crop plants used as food. On splitting, they produce dal which are rich in protein such as green gram, black gram, pea, cowpea, pigeonpea, soybean, lentil.
- Feed and forage crops: It refers to crops used as feed for animals. It includes crops like berseem, cow-pea, field bean, guinea grass, lucerne, maize, napier, oats, para grass, sudan grass, turnips and velvet bean and hay, silage and fodder obtained form the said crops.
- Fibre crop: Crop plants that are grown for fibre yield. Fibre is obtained from crops, such as cotton, jute, mesta, roselle, sunnhemp and flax.
- Sugar crop: Crop grown for the production of sugar are sugarcane and sugar beet.
- Spices and condiments: Crop plants or their products are used to season, flavour, taste and add zest and sometimes colour the fresh or preserved food; for instance ginger, garlic, fenugreek, cumin, turmeric, chilli, onion, coriander.
- Drug crop: Crop that is used for the preparation of medicines, for instance tobacco, mint.
- Narcotics, fumitories and masticatories: Crop plants or their products that are used for stimulating, numbing, drowsing or relishing effects such as tobacco, ganja, opium, poppy, anise.
- Beverage: Products of crops used for mild, agreeable and stimulating liquors meant for drinking such as tea, coffee and cocoa.
- Fruit crop: Jack fruit, mango, pine apple, banana etc are major fruit crops of the country.
- Special purpose crop/ industrial crop: Mulberry is the main industrial crop of the country grown only for silk worm and silk production.

Cash Crop

A cash crop or profit crop is an agricultural crop which is grown to sell for profit. It is typically purchased by parties separate from a farm. The term is used to differentiate marketed crops from subsistence crops, which are those fed to the producer's own livestock or grown as food for the producer's family. In earlier times cash crops were usually only a small (but vital) part of a farm's total yield, while today, especially in developed countries, almost all crops are mainly grown for revenue. In the least developed countries, cash crops are usually crops which attract demand in more developed nations, and hence have some export value.

Prices for major cash crops are set in commodity markets with global scope, with some local variation (termed as "basis") based on freight costs and local supply and demand balance. A consequence of this is that a nation, region, or individual producer relying on such a crop may suffer low prices should a bumper crop elsewhere lead to excess supply on the global markets. This system has been criticized by traditional farmers. Coffee is an example of a product that has been susceptible to significant commodity futures price variations.

Globalization

Issues involving subsidies and trade barriers on such crops have become controversial in discussions of globalization. Many developing countries take the position that the current international trade system is unfair because it has caused tariffs to be lowered in industrial goods while allowing for low tariffs and agricultural subsidies for agricultural goods. This makes it difficult for a developing nation to export its goods overseas, and forces developing nations to compete with imported goods which are exported from developed nations at artificially low prices. The practice of exporting at artificially low prices is known as dumping, and is illegal in most nations. Controversy over this issue led to the collapse of the Cancún trade talks in 2003, when the Group of 22 refused to consider agenda items proposed by the European Union unless the issue of agricultural subsidies was addressed.

Per Climate Zones

Arctic

The Arctic climate is generally not conducive for the cultivation of cash crops. However, one potential cash crop for the Arctic is *Rhodiola rosea*, a hardy plant used as a medicinal herb that grows in the Arctic. There is currently consumer demand for the plant, but the available supply is less than the demand.

Temperate

A tea plantation in the Cameron Highlands in Malaysia.

Cash crops grown in regions with a temperate climate include many cereals (wheat, rye, corn, barley, oats), oil-yielding crops (e.g. grapeseed, mustard seeds), vegetables (e.g.

potatoes), tree fruit or top fruit (e.g. apples, cherries) and soft fruit (e.g. strawberries, raspberries).

Subtropical

In regions with a subtropical climate, oil-yielding crops (e.g. soybeans) and some vegetables and herbs are the predominant cash crops.

Tropical

In regions with a tropical climate, coffee, cocoa, sugar cane, bananas, oranges, cotton and jute (a soft, shiny vegetable fiber that can be spun into coarse, strong threads), are common cash crops. The oil palm is a tropical palm tree, and the fruit from it is used to make palm oil.

Global Cash Crops

Coconut palms are cultivated in more than 80 countries of the world, with a total production of 61 million tonnes per year. The oil and milk derived from it are commonly used in cooking and frying; coconut oil is also widely used in soaps and cosmetics.

Sustainability of Cash Crops

Approximately 70% of the world's food is produced by 500 million smallholder farmers. For their livelihood they depend on the production of cash crops, basic commodities that are hard to differentiate in the market. The great majority (80%) of the world's farms measure 2 hectares or less. These smallholder farmers are mainly found in developing countries and are often unorganized, illiterate or enjoyed only basic education. Smallholder farmers have little bargaining power and incomes are low, leading to a situation in which they cannot invest much in upscaling their businesses. In general, farmers lack access to agricultural inputs and finance, and do not have enough knowledge on good agricultural and business practices. These high level problems are in many cases threatening the future of agricultural sectors and theories start evolving on how to secure a sustainable future for agriculture. Sustainable market transformations are initiated in which industry leaders work together in a pre-competitive environment to change market conditions. Sustainable intensification focuses on facilitating entrepreneurial farmers. To stimulate farm investment, projects on access to finance for agriculture are also popping up. One example is the SCOPE methodology, an assessment tool that measures the management maturity and professionalism of producer organizations as to give financing organizations better insights in the risks involved in financing. Currently agricultural finance is always considered risky and avoided by financial institutions.

Black Market Cash Crops

In the U.S., cannabis has been termed as a cash crop.

Coca, opium poppies and cannabis are significant black market cash crops, the prevalence of which varies. In the United States, cannabis is considered by some to be the most valuable cash crop. In 2006, it was reported in a study by Jon Gettman, a marijuana policy researcher, that in contrast to government figures for legal crops such as corn and wheat and using the study's projections for U.S. cannabis production at that time, cannabis was cited as "the top cash crop in 12 states and among the top three cash crops in 30". The study also estimated cannabis production at the time (in 2006) to be valued at $35.8 billion USD, which exceeded the combined value of corn at $23.3 billion and wheat at $7.5 billion.

Catch Crop

Catch crop is a fast-growing crop that is grown between successive plantings of a main crop.

For example, radishes that mature from seed in 25–30 days can be grown between rows of most vegetables, and harvested long before the main crop matures. Or, a catch crop can be planted between the spring harvest and fall planting of some crops.

Catch cropping is a type of succession planting. It makes more efficient use of growing space.

Catch crops are also crops that are sown to prevent minerals being flushed away from the soil: crops such as grain are used to keep certain minerals not attached to the soil-clay connection in the soil.

Advantages of Catch Crops

- Reducing leaching,
- Reducing erosion,
- Weed and volunteer crop suppression,

- Increasing biological activity in the soil,
- Adding organic matter to the soil,
- Improving soil structure,
- Root penetration of various soil horizons,
- Potential to reduce pests and disease,
- Increasing the supply of nutrients for following crops,
- Legumes add extra Nitrogen to the soil,
- Potential to produce animal feed for use in winter or spring.

Disadvantages of Catch Crops

- May increase certain pests and diseases (ensure correct crop choice to minimise risk),
- Potential for catch crop species to become weeds in future crops (do not allow catch crop plants to set seed),
- Pressures of workload at optimum establishment time,
- Successful establishment is difficult in very dry weather,
- Cost of establishing and incorporating catch crops.

Cereals

Cereal, also called grain is any grass (family Poaceae) yielding starchy seeds suitable for food. Most grains have similar dietary properties; they are rich in carbohydrates but comparatively low in protein and naturally deficient in calcium and vitamin A. Breads, especially those made with refined flours, are usually enriched in order to compensate for any nutritional deficiencies in the cereal used. The cereals most commonly cultivated are wheat, rice, rye, oats, barley, corn (maize), and sorghum.

As human food, cereals are usually marketed in their raw grain form (some are frozen or canned) or as ingredients of various food products. As animal feed, they are consumed mainly by livestock and poultry, which are eventually rendered as meat, dairy, and poultry products for human consumption. Many cereals are used industrially in the production of a wide range of substances, such as glucose, adhesives, oils, and alcohols.

Corn, or maize (Zea mays), was originally domesticated in the Western Hemisphere by Native Americans and was then carried to Europe by the early explorers. It is a major crop cultivated in most temperate climates, although the United States is the single largest producer. For human consumption, corn is sold as a fresh vegetable or is canned or frozen. The grain also is processed into a growing number of food products, including corn flour, corn oil, corn syrup, and many other by-products. It is a very important animal feed and is heavily used in the production of cellulosic ethanol, a biofuel.

Cornfield.

Rice (Oryza sativa) is the second largest cereal crop and is a staple food in all areas of Asia. Unlike wheat, which is generally raised on large farms and harvested mechanically, rice is usually grown on small paddies and harvested by hand. Cultivation methods have changed little over the centuries; the paddies are inundated with water, usually up to about 15 cm (6 inches), then drained and dried just before harvest. Most rice is milled for direct local consumption. Other products in which rice is used are breakfast cereals and such alcoholic beverages as Japanese sake.

Wheat (various Triticum species) is a major cereal crop and one of the oldest domesticated grains. In modern times, wheat is used to produce meal, breakfast cereals, and flour for bakery products. It can be cultivated in a wide range of soils but thrives in temperate climates.

Rye (Secale cereale) is widely used for bread making, second only to wheat for that purpose. It is also used in other bakery products and in distilled liquors, especially whiskey. Rye can be grown on relatively poor soils and is able to survive more severe winters than most grains. Poland and Russia are some of the world's largest rye producers.

Dark rye bread.

Oats (Avena sativa) are grown in most of the temperate regions of the world, especially in the United States, Canada, and northern Europe. Most of the oats produced are used in animal feed, although they may also be processed for human consumption.

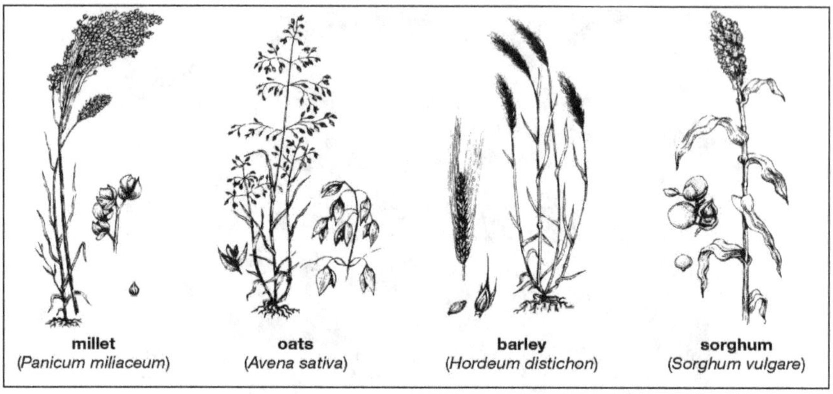

Grain.

Barley (Hordeum vulgare) is also grown in temperate climates. It does not need especially rich soils. Most barley is used for animal feed, though it is also the source of the malt used in both the brewing of beer and the distillation of alcoholic beverages. Barley malt is also a constituent of vinegar and many breakfast foods.

Sorghum (Sorghum vulgare), also called milo, is principally grown for use as animal feed. Teff (Eragrostis tef) and millet (various species) are locally grown in a number of countries both for human consumption and as livestock feed.

Food Crops

The term "food crops" refers to the world's major food supply derived from plants; a crop assumes human intervention through agriculture. In the main, food crops consist of grains, legumes (including dried beans), seeds and nuts, vegetables, fruit, herbs and spices, beverage plants such as tea and coffee, and so forth.

Some foods, like sea vegetables, are food crops in terms of gathering rather than deliberate agriculture, although undoubtedly the best sources of such foods are well cared for resources in terms of laws to prevent over-harvesting, polluting, and licensing access.

Grains

Grains are a principal source of human food, with wheat, rice, barley, oats, millet, rye and corn (maize) being the most widely consumed sources of grains around the world. Grains are the seeds of grasses. Buckwheat is mistaken for a grain, but is the seed of a herbaceous plant, not a grass.

Wheat, millet and rice are the richest source of proteins. Grains have served as a staple food for much of the world's population for thousands of years. In particular, the staples of rye, wheat, barley and oats have been temperate climate favourites, whilst rice, corn and millet have been favored in the tropics and sub-tropics. In modern times, affluent countries are able to enjoy a very wide range of grains that once used to be very localised. However, affluent countries have also taken to using grain to feed livestock to produce meat and less grains are eaten at the dinner table as staple produce as a result.

Dried Beans and Legumes

These are another staple in many diets around the world. Generally cheap to buy, dried beans and legumes tend to also have a long shelf life, making them good for seasons when food is more scarce. They are a high quality source of protein, although on their own they tend to be less tasty and need to be combined with other foods to make them palatable. Besides protein, they contain fibre, carbohydrate, minerals and vitamins needed by the human body. They don't have vitamin C unless sprouted and eaten as sprouts.

One well known food made from soybeans is tofu or bean curd, which is then used in thousands of different ways in varied dishes. Although tasteless, it takes on other flavours readily, and its texture makes it ideal to add in place of meat in many dishes.

Seeds and Nuts

The variety of seeds and nuts available as food crops is enormous. Both seeds and nuts are botanically the same because they consist of a kernel that contains a future plant in embryonic form. As such, they are a powerhouse of energy and a concentrated form of food.

Vegetables

Vegetables are an excellent source of vitamins, minerals and roughage in the human diet. Leaves are particularly important in terms of nutrients, while root vegetables contain the starch and natural sugars that supply humans with energy.

Vegetables are grown both commercially and as a typical backyard or sustainable lifestyle food crop. Organic vegetables tend to cost more and tend to have higher nutritional content. Some say they also taste better than conventionally grown vegetable crops. not all naturally grown foods are vegetables.

Fruit

Fruit tends to be grown in orchards, berry patches, on vines and even in bogs (for cranberries). Most fruit has a specific season and many fruits spoil quickly after picking, so fast action to get them to market or into refrigeration is essential. Wild sources of fruit are not usually considered to be a food crop, although established seasons of collecting berries from areas known for bountiful crops can become a traditional source of berries in a localised area.

Most fruit contains vitamin C, making it an important source of vitamin C in humans.

Dried fruit is another common source of fruit in the diet and parts of fruit crops are often set aside for this purpose.

Fruit tends to be grown commercially and in home gardens, although fruit trees can take many years to be sufficiently mature to produce adequate fruit for a household.

Herbs and Spices

Herbs are mostly annuals but there are biannuals and perennials too. Herbs consist mainly of the best leaves, stems, flowers, seeds and roots. Some herbs are very strong in flavour. It is commonplace to grow fresh herbs for sale in supermarkets in affluent countries, as an option for a kitchen garden and live, fresh herbs to cook with.

Spices consist of the dried roots, bark, pods, berries or seeds of aromatic plants. Examples include nutmeg, pepper, cinnamon and saffron. Much of the history of trade is bound up in the transfer of spices from East to West, as well as the discovery of vanilla, chillies and allspice in the New World.

Energy Crop

Energy crops are low-cost and low-maintenance crops grown solely for energy production (not for food). The crops are processed into solid, liquid or gaseous fuels, such as pellets, bioethanol or biogas. The fuels are burned to generate power or heat.

The plants are generally categorized as woody or herbaceous. Woody plants include willow and poplar, herbaceous plants include *Miscanthus x giganteus* and *Pennisetum purpureum* (both known as elephant grass). Herbaceous crops, while physically smaller than trees, store roughly twice the amount of CO_2 (in the form of carbon) below ground, compared to woody crops.

Through biotechnological procedures such as genetic modification plants can be manipulated to create higher yields. Relatively high yields can also be realized with existing cultivars. However, some additional advantages such as reduced associated costs (i.e. costs during the manufacturing process) and less water use can only be accomplished by using genetically modified crops.

Relationship between above-ground yield (diagonal lines), soil organic carbon (X axis), and soil's potential for successful/unsuccessful carbon sequestration (Y axis). Basically, the higher the yield, the more land is usable as a GHG mitigation tool (including relatively carbon rich land).

The amount of carbon sequestrated and the amount of GHG (greenhouse gases) emitted will determine if the total GHG life cycle cost of a bio-energy project is positive, neutral or negative. Specifically, a GHG/carbon negative life cycle is possible if the total below-ground carbon accumulation more than compensates for the above-ground total life-cycle GHG emissions. Whitaker et al. estimates that for Miscanthus x giganteus, carbon neutrality and even negativity is within reach. Basically, the yield and related carbon sequestration is so high that it more than compensates for both farm operations emissions, fuel conversion emissions and transport emissions. The graphic on the right displays two CO_2 negative Miscanthus x giganteus production pathways, represented in gram CO_2-equivalents per megajoule. The yellow diamonds represent mean values.

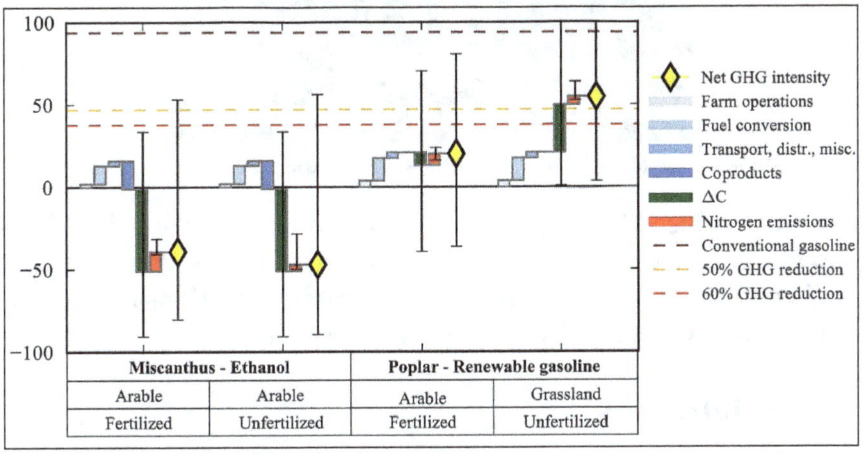

GHG/CO_2/carbon negativity for Miscanthus x giganteus production pathways.

One should note that successful sequestration is dependent on planting sites, as the best soils for sequestration are those that are currently low in carbon. The varied results displayed in the graph highlights this fact. Milner et al. argues that for the UK, successful sequestration is expected for arable land over most of England and Wales, with unsuccessful sequestration expected in parts of Scotland, due to already carbon rich soils (existing woodland). Also, for Scotland, the relatively lower yields in this colder climate makes CO_2 negativity harder to achieve. Soils already rich in carbon includes peatland and mature forest. Grassland can also be carbon rich, however Milner et al. further argues that the most successful carbon sequestration in the UK takes place below improved grasslands. The bottom graphic displays the estimated yield necessary to achieve CO_2 negativity for different levels of existing soil carbon saturation.

The perennial rather than annual nature of Miscanthus crops implies that the significant below-ground carbon accumulation each year is allowed to continue undisturbed. No annual plowing or digging means no increased carbon oxidation and no stimulation of the microbe populations in the soil, and therefore no accelerated carbon-to-CO_2 conversion happening in the soil every spring.

Types

Solid Biomass

Solid biomass, often pelletized, is used for combustion in thermal power stations, either alone or co-fired with other fuels. Alternatively it may be used for heat or combined heat and power (CHP) production.

Elephant grass (Miscanthus giganteus) is an experimental energy crop.

In short rotation coppice (SRC) agriculture, fast growing tree species like willow and poplar are grown and harvested in short cycles of three to five years. These trees grow best in wet soil conditions. An influence on local water conditions can not be excluded. Establishment close to vulnerable wetland should be avoided.

Gas Biomass (Methane)

Whole crops such as maize, Sudan grass, millet, white sweet clover, and many others can be made into silage and then converted into biogas. Anaerobic digesters or biogas plants can be directly supplemented with energy crops once they have been ensiled into silage. The fastest growing sector of German biofarming has been in the area of "Renewable Energy Crops" on nearly 500,000 ha (1,200,000 acres) of land. Energy crops can also be grown to boost gas yields where feedstocks have a low energy content, such as manures and spoiled grain. It is estimated that the energy yield presently of bioenergy crops converted via silage to methane is about 2 GWh/km^2 (1.8×10^{10} BTU/sq mi) annually. Small mixed cropping enterprises with animals can use a portion of their acreage to grow and convert energy crops and sustain the entire farms energy requirements with about one fifth of the acreage. In Europe and especially Germany, however, this rapid growth has occurred only with substantial government support, as in the German bonus system for renewable energy. Similar developments of integrating crop farming and bioenergy production via silage-methane have been almost entirely overlooked in N. America, where political and structural issues and a huge continued push to centralize energy production has overshadowed positive developments.

Liquid Biomass

Biodiesel

Coconuts sun-dried in Kozhikode, Kerala for making copra, the dried meat, or kernel, of the coconut. Coconut oil extracted from it has made copra an important agricultural commodity for many coconut-producing countries. It also yields coconut cake which is mainly used as feed for livestock.

Coconuts sun-dried in Kozhikode, Kerala.

Pure biodiesel (B-100), made from soybeans.

European production of biodiesel from energy crops has grown steadily in the last decade, principally focused on rapeseed used for oil and energy. Production of oil/biodiesel from rape covers more than 12,000 km² in Germany alone, and has doubled in the past 15 years. Typical yield of oil as pure biodiesel may be is 100,000 L/km² (68,000 US gal/sq mi; 57,000 imp gal/sq mi) or more, making biodiesel crops economically attractive, provided sustainable crop rotations exist that are nutrient-balanced and preventative of the spread of disease such as clubroot. Biodiesel yield of soybeans is significantly lower than that of rape.

Typical oil extractable by weight	
Crop	Oil %
Copra	62
Castor seed	50
Sesame	50
Groundnut kernel	42
Jatropha	40
Rapeseed	37
Palm kernel	36
Mustard seed	35
Sunflower	32
Palm fruit	20
Soybean	14
Cotton seed	13

Bioethanol

Energy crops for biobutanol are grasses. Two leading non-food crops for the production of cellulosic bioethanol are switchgrass and giant miscanthus. There has been a preoccupation with cellulosic bioethanol in America as the agricultural structure supporting biomethane is absent in many regions, with no credits or bonus system in place. Consequently, a lot of private money and investor hopes are being pinned on marketable and patentable innovations in enzyme hydrolysis and the like.

Bioethanol also refers to the technology of using principally corn (maize seed) to make ethanol directly through fermentation, a process that under certain field and process conditions can consume as much energy as is the energy value of the ethanol it produces, therefore being non-sustainable. New developments in converting grain stillage (referred to as distillers grain stillage or DGS) into biogas energy looks promising as a means to improve the poor energy ratio of this type of bioethanol process.

NEGLECTED AND UNDERUTILIZED CROP

Neglected and underused crops are domesticated plant species that have been used for centuries or more for their food, fibre, fodder, oil or medicinal properties, but have been reduced in importance over time. A wide range of terminology is used to describe these crops, including: orphan, abandoned, lost, underutilized, local, minor, traditional, alternative, niche, or underdeveloped; in recent fashion, these are often referred to as forgotten or smart food. Reductions in use may pertain to supply or consumption constraints. These can include, *inter alia*, poor shelf life, unrecognized nutritional value, poor consumer awareness and reputational problems (famine food or "poor people's food", sometimes due to the modernization of agricultural practices). Some crops have been so neglected that genetic erosion of their genepools has become so severe that they are often regarded as lost crops.

Cherimoya (Annona cherimola) on sale in Cali, Colombia. In left background: domestically produced mangosteen (Garcinia mangostana).

As the demand for plant and crop attributes changes (reappraisal or discovery of nutritional traits, culinary value, adaptation to climate change, etc.), neglected crops can overcome the constraints to the wider production and use. As a matter of fact, many formerly neglected crops are now globally significant crops (oilpalm, soybean, kiwi fruit). Although the options for scaling up neglected crops for large-scale agriculture appear to be increasingly exhausted, many species have the potential to contribute to food security, nutrition, dietary and culinary diversification, health and income generation. They also provide environmental services. It is impossible to define what would constitute "proper" or "correct" levels of use; however, many neglected species evidently are underused relative to their nutritional value and productivity.

Bambara groundnut (Vigna subterranea) from Buzi district in Mozambique.

Adding to that, orphan crops also helps in food security. Thus exist when all people at all times have physical and economic access to sufficient, safe, and nutritious food to meet their dietary needs and food preferences for an active and healthy life, especially during famine.

Landscape with quinoa (Chenopodium quinoa), Cachilaya, Bolivia, Province La Paz, Lake Titicaca seen in background.

Just three crops - maize, wheat and rice - account for about 50% of the world's consumption of calories and protein. About 95% of the world's food needs are provided by just 30 species of plants. In stark contrast, at least 12,650 species names have been compiled as edible. Neglected and underutilized plants are those that could be - and, in many cases, historically have been - used for food and other uses on a larger scale.

Such crop species have also been described as "minor", "orphan", "promising" and "little-used".

It is difficult to precisely define which attributes make a crop "underutilized", but often they display the following features:

- Linkage with the cultural heritage of their places of origin,
- Local and traditional crops whose distribution, biology, cultivation and uses are poorly documented,
- Adaptation to specific agroecological niches and marginal land,
- Weak or no formal seed supply systems,
- Traditional uses in localized areas,
- Produced in traditional production systems with little or no external inputs,
- Receive little attention from research, extension services, policy and decision makers, donors, technology providers and consumers,
- May be highly nutritious and/or have medicinal properties or other multiple uses.

Neglected crops are primarily grown by traditional farmers. These species may be widely distributed beyond their centres of origin but tend to occupy special niches in the local production and consumption systems. They are important for the subsistence of local communities, yet remain poorly documented and neglected by the mainstream research and development activities.

Underutilized crops are those marginalized by farmers and consumers due to agronomic, genetic, economic, environmental and cultural reason, which were once important and major crop in the community.

Importance

They continue to play an important role in the subsistence and economy of poor people throughout the developing world, particularly in the agrobiodiversity-rich tropics. Despite their potential for dietary diversification and the provision of micronutrients such as vitamins and minerals, they continue to attract little research and development attention.

Alongside their commercial potential, many of the underused crops also provide important environmental services, as they are adapted to marginal soil and climate conditions.

Constraints

The following are frequent constraints:

- Limited germplasm available;
- Lack of technical information;
- Lack of national policy;
- Lack of interest by researchers, agriculturists and extension workers;
- Lack of producer interest.

Village chief of Boula Téné, Senegal. Theodore Mada Keita, holds up the fonio grain, White fonio (Digitaria exilis) that helps feed his family in southern Senegal.

Determination of the underutilized status of a crop varies among researchers. Different criteria and approaches are used to define this particular group of crop. Neglect refers to the attention the crop has received from research and development and can be evaluated by how well national and international policy and legal frameworks and research and development programmes support the conservation and sustainable use of the crop. Underutilisation is particular to the geography and potential for a crop to contribute to better to diets and production systems. In any cases where exotic species or diversified species are underutilized at certain region, these are not necessarily underutilized in other parts of the worlds.

CROP SCIENCE

Crop science is the study of the world's major food, feed, turf, and fiber crops and their environment. It is a broad discipline encompassing breeding, genetics, production, and management. Crop scientists work in a variety of topic areas and professional settings.

They all work toward the same goal: making our food healthier and more sustainable for a growing population.

Crop physiologists research how plants grow and function.

Some crop physiologists study individual plant cells. Others specialize in specific plant parts like leaves, stems, or roots. And yet others study the interactions of plants in their environments. Crop physiologists work to gain knowledge about what controls plant growth and how plants respond to their changing environment.

Plants make life on earth possible because they convert solar energy into chemical energy we can use. That process is photosynthesis. We all depend on plant products for food, fiber, medicines, and a wide array of natural chemicals. How do they grow? What limitations do they face? Which plants are more efficient? Crop physiologists fill the knowledge gap between processes controlling growth and seed development and the targeted production of economically-valuable plant products.

Challenges

1. Climate change: Our major crop plants remain vulnerable to drought. We understand many of the mechanisms involved and have been breeding for plants that are more drought-tolerant for decades. But the climate is changing. Weather patterns are becoming more variable. At the same time, demands for food, fiber, and energy production continue to rise. Crop scientists must find plants that can help meet these demands.

These two soybean varieties show a visible difference in how quickly they mature. By finding the genetic reason for this difference, crop scientists can improve growing habits of soy and other crops.

2. More efficient water and nutrient use: Our major crop plants use 250 to 530 gallons of water for every two pounds of grain they produce. (That's about 20 laundry machine loads worth of water!) By understanding how plants use water, we may be able to breed plants that are more efficient. Plants also get most of their nutrients from the soil. How their root systems interact with the soil environment may hold a key to breeding plants that are more efficient in nutrient use.

3. Lack of new cropland: The amount of land available for growing food and fiber is decreasing as the human population increases. We have to use the farmland we have more efficiently. Crop physiologists are working to understand why some plants are more efficient, producing more per acre. By growing the crops that are the most efficient at using water, nutrients, and the sun's energy, we can harvest more—and harvest crops that are more nutritious.

4. Soil biology: Knowledge of plant interactions with favorable soil microorganisms is very limited. Soil microbes can help plants get nutrients, help defend against diseases, and even create growth-promoting compounds. Studying how this happens may lead to discovering crop plants that work more closely with the microbes in the soil. This can reduce the level of fertilizers and pesticides farmers need to apply while still increasing yields.

Advancements in Crop Biology

These corn plants are growing in a waterlogged claypan soil after a heavy rainfall and strong winds. Develping crops that are resilient to such extreme weather is a critical goal of crop scientists.

Discoveries by crop physiologists have had major impacts on our knowledge of plant life, their uses, and agricultural practices. Here are a few that are particularly noteworthy:

- How plants receive the sun's rays and their ability to use it for greater grain yield,
- Basic plant mechanisms that increase tolerance to drought and flooding,
- Plant hormones and how they work,
- How a plant's flowering time frame is changed by temperature and daylight,
- How plants use carbon and nitrogen,
- How plants transport water and other nutrients within them,
- How plants "decide" which parts need most growth,

- How the environment changes seeds and other plant products,
- How plants and microbes interact.

Crop Sustainability

Just like forests and deserts, farms have unique ways of functioning. Crop ecologists study how agricultural plant communities respond to their surrounding environment. This includes looking at larger ecosystems, such as:

- Plants' physical environment,
- Weather,
- Soil,
- Minerals and other elements present,
- The biological needs of the crops and their neighboring plants, and
- Fertilizers added.

The field environments that crops grow in are called agroecosystems. By studying these unique environments, scientists can develop ways to improve farmers' incomes and preserve natural resources. To do this, crop ecologists focus on innovative and sustainable farming systems.

Challenges

Meeting food demand while minimizing our environmental footprint is one of the greatest challenges humanity has ever faced. There is less arable land—land good for growing crops. And yet, the world population is growing. Leaders predict global food production must increase by at least 70% by 2050 to support current population growth trends. Water scarcity is also a challenge for crop ecologists. Creating agroecosystems that produce more food with existing—or less—land and water is a top goal.

Advancements in Crop Sustainability

A grower's field can show wide variability. The different shades of green show differences in soil quality, moisture, and nutrients that affect the crop's growth.

Average farm yields for major crops like corn, wheat, rice, and soybeans have increased greatly over the past 50 years. Crop ecologists have played a role in this success! They examined plant genetics to find the best varieties. They researched environmental factors. And they promoted improved farm practices, or management, that contributed to these gains. These insights are the foundation of continuing research. Scientists continue to develop new cultivars, identify the best management practices, and investigate what holds back crop yields.

Crop ecologists are now applying a strategy called site-specific management. Farmland varies in soil fertility, topography, and yield potential. Techniques used on one farm may not work on another. In fact, what works in one area of one farm might not work elsewhere on the same farm. Site-specific management means seeds, water, and nutrients are used in a way that is best for that exact piece of land. Precision farming machinery, equipped with satellite or GPS guidance, offers a high-tech solution. The result is more efficient and sustainable treatment of farmland. There is less waste and a better yield. In this way site-specific management helps reduce the environmental footprint of modern farming operations.

SEED TECHNOLOGY

Seed sciences and technologies process, test, and conserve food and agricultural seeds. The goal is to improve the quality of the seeds that farmers plant, and the food plants that people and livestock eat.

When you hold a seed in your hand, no matter how small it is, you are holding a future plant! The plant's blueprint is encapsulated, waiting for moisture and nutrients to press the "go" button. Seeds are amazing.

Preparing seeds for storage.

Seeds make up over 70% of the food that people consume directly. From the wheat in our bread to the oats in our cereal, from corn and rice to beans and peas, seeds nourish humankind. Seeds are also planted to produce the rest of the plant foods we eat. From the leaves of lettuce to the fruits of tomatoes, it all starts with a seed. When we eat meat, dairy, and eggs, we also rely on seeds and their products, as livestock are largely fed on corn, soybean, and other seeds. The fibers that clothe us, such as cotton, linen, and hemp, are grown from seeds.

Food and agricultural seeds are developed, processed, and maintained with high quality through continuous advancements in seed sciences and technologies. Some of the steps scientists take to assure a safe seed supply are:

- Sorting
- Drying
- Testing
- Cleaning
- Packaging
- Storing

Continuous advancements in seed sciences and technologies ensure seeds are developed, processed, and maintained with high quality. If a seed supply is not cared for properly, it will lose its ability to sprout, resulting in empty fields. Seed security is food security.

Challenges

These sorghum seeds will be dependable for growers.

Productive crops and healthy plant foods for people and livestock start with good seeds. Seed producers and processors must provide these high-quality seeds to farmers and consumers every year. This is challenging work, as it requires that seeds be clean and free of weed seeds and other contaminants. Seeds for sale must also be certified for their viability and may be treated in preparation for planting. They need to be checked to make sure they have a good germination rate. Seeds conserved as genetic resources can be used in future plant breeding efforts must be preserved with care so that they are available for decades or longer. These high standards require that seed processing and testing is performed both rapidly and with precision. Seed sciences and technologies support and improve these processes.

NUTRITIONAL QUALITY IN CROP SCIENCE

Plants provide a major part of a nutritious diet. Fruits and vegetables are harvested directly from plants. They are important sources of minerals, vitamins, fiber, and energy. Many common foods like bread, cereals, and pasta are processed from ingredients that come from plants. Even our animal-based foods such as meat, eggs, dairy, and fish depend on plants. Understanding how plants collect, produce, and store health-beneficial nutrients is an important goal of plant scientists.

Plant scientists have made tremendous advances understanding what makes plant products nutritious. The roles of vitamins, minerals, and other compounds found in plants are becoming clear. Plant breeders have been improving crop plants for thousands of years, making them more productive, nutritious, and easier to grow and harvest. Recently, plant scientists have found ways to make plants even more nutritious to help those suffering from nutrient deficiencies.

Challenges

A colorful variety of vegetables provides a wider assortment of nutrients.

Many people in poor countries lack the vitamins or minerals needed for good health. In addition, the foods they do eat often come from nutrient-deficient soils. If the soils are nutrient-deficient, so are the foods. About two billion people around the world suffer from hidden hunger: getting enough calories, but not enough nutrition. Blindness, cognitive impairments, stunting, and premature death can result.

There are a few ways to solve this malnutrition problem. One is to provide vitamin supplements. Another is to fortify the foods. You may know, for example, that the bread and cereals you buy at the grocery store are fortified. In these examples, food production companies add vitamins and minerals to these products.

Scientists worked with a team of researchers to develop food that actually had more nutrition inside the plant. Other scientists worried that this would hurt the plant's yields, but the team discovered that they could increase both zinc content and yield. Zinc is essential for a healthy immune system, growth, wound healing, reproduction, fertility, and sense of taste and smell.

The process of growing foods with increased nutrition is called biofortification.

Another successful example is orange-fleshed sweet potatoes. White-fleshed sweet potato is an important crop in some African countries. It provides more edible energy per meal than wheat, rice, or cassava. Orange-fleshed sweet potatoes are derived from traditional white sweet potatoes but contain more beta-carotene. The deep orange color is the result. Our bodies convert beta-carotene into vitamin A. Many people in undeveloped or under-developed countries have vitamin A deficiencies. This can lead to blindness and other preventable diseases. Eating biofortified foods like orange-flesh sweet potatoes helps to overcome these deficiencies. The result is a healthier, more productive society.

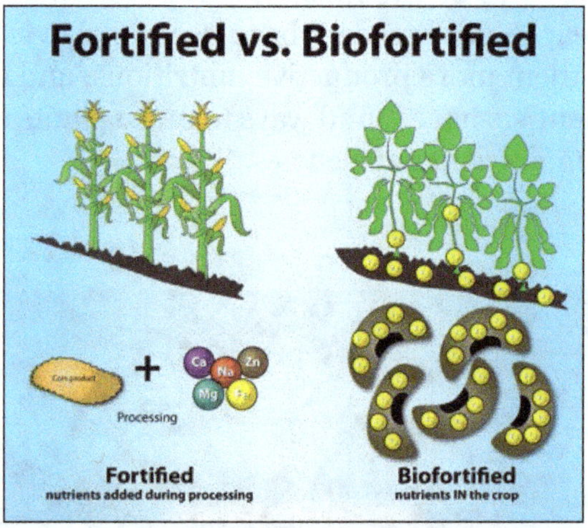

Fortified foods have nutrients added during processing. Biofortified crops are bred to take in more nutrients from the soil – and make the crops more nutritious.

Improving the nutritional value of plants and seeds might also involve decreasing levels of naturally occurring anti-nutritionals or toxins. A few examples plant scientists have developed include decreased glucosinolates in oilseed rape, decreased unsaturated oils and trypsin inhibitors in soybean, and low phytate levels in cereals.

Crop physiologists and molecular biologists help breeders develop crop varieties with improved nutritional value and lower levels of undesirable compounds. Together they uncover the genes and the biochemical pathways that contribute to these advances. New technologies such as genome editing will undoubtedly allow development of new types of nutritionally enhanced plants.

GRAZING AND FORAGE SCIENCE

Forage and grassland scientists study plants eaten by livestock. This might include domesticated cattle, horses, goats, and sheep, or even wild buffalo. Today's forage scientist also looks at creating grasslands that provide habitats for wildlife. They research grasses that encourage conservation of natural resources. An emerging field in grassland science is bioenergy. Grasslands create diverse, perennial ecosystems, so research is ongoing for their environmental benefits.

How does this Affect?

Much of today's agricultural land produces grains, vegetables, and fruits eaten by humans. But some land can't support these types of crops due to soil quality and other issues. This land could be used effectively for forage or grasslands.

Although grasses are often unsuitable for human consumption, animals that eat these crops are able to convert them into valuable protein (meat) and fiber (for example, wool). Grasslands support a wide range of wild animals, birds, insects, and plants. They even provide a space for a broad range of human recreational activities such as bird-watching, hiking, hunting, and wildlife observation.

Research has shown that grasslands can also provide environmental protection. Grasses planted at the outer edge of agricultural land can trap nutrients that might otherwise run off into neighboring streams. Growers and researchers refer to these naturalized areas on farms as buffer strips.

Challenges

Good land managers consider the full picture of range management.
This Montana ranch uses rotational grazing.

Climate change has a large impact on forage and grassland systems. Violent storms with large downpours cause flooding. In other years, these crops are subjected to long periods of drought. Breeding grass cultivars that are heat and drought resilient is important. Many grasses have long roots that can reach down several feet into the soil to pull up nutrients and water. Scientists are still studying the underlying physiology and biochemistry of forage plants that influence yield, quality, and persistence. Research is yielding valuable and interesting information about grassland systems.

What Advances have we Seen in Grazing and Forage?

Over the last century, plant breeders have made great progress in developing plants with desirable traits.

Today's alfalfa can grow under a broader range of environmental conditions with better yields. Alfalfa is an important crop because it is a nitrogen fixing plant: it can pull nitrogen from the air, transform it, and use it as a nutrient. This trait makes it a good plant to improve soil health.

Tall fescue is important because it is a very hardy and nutritious forage crop. Tall fescue is a grass that lives in harmony with various fungi. The grass provides a home for the fungi while the fungi provide nutrients to the grass. This is great for the tall fescue! However, one type of fungus that works well with grasses can make cattle sick. Researchers have studied other fungi that cattle could tolerate. As a result, plant breeders developed newer types of tall fescue that grow with fungi that help the grass and don't harm the cattle.

Forage grasses have also been bred to be more nutritious. Forage and grassland scientists worked closely with animal scientists to develop grazing and hay production systems that maximize animal productivity while maintaining the persistence and productivity of grassland systems.

References

- Definition-of-crop-and-agricultural: blogspot.com, Retrieved 12 July, 2019
- USDA-Foreign Agriculture Service. "(Cotton) Production Ranking MY 2011". National Cotton Council of America. Retrieved April 3, 2012
- Grazing-forage about-crop-science: crops.org, Retrieved 15 June, 2019
- Hillstrom, Kevin; Hillstrom, Laurie Collier (2003). Australia, Oceania, and Antarctica: A Continental Overview of Environmental Issues. ABC-CLIO, Inc. Pp. 73–119. Retrieved April 9, 2012. ISBN 1576076954
- Catch-crops, arable-seeds, agriculture: goldcrop.ie, Retrieved 1 August, 2019
- J. A. R. Lockhart; A. J. L. Wiseman (17 May 2014). Introduction to crop husbandry including grassland). Elsevier. P. 111. Retrieved 24 June 2015
- Food-crops: appropedia.org, Retrieved 5 March, 2019

- Ara Kirakosyan; Peter B. Kaufman (2009-08-15). Recent Advances in Plant Biotechnology. P. 169. ISBN 9781441901934. Retrieved 14 February 2013
- Cereal, topic: britannica.com, Retrieved 11 January, 2019
- Hartwich, Jens; Bölscher, Jens; Schulte, Achim (2014-09-19). "Impact of short-rotation coppice on water and land resources". Water International. 39 (6): 813–825. Doi:10.1080/02508060.2014.959870. ISSN 0250-8060
- Seed-technology, about-crop-science: crops.org, Retrieved 3 May, 2019
- O'Malley, Brendan (26 November 2008). "New centre to boost knowledge on underutilised crops". Scidev.Net. Retrieved 21 August 2013
- Nutritional-quality about-crop-science: crops.org, Retrieved 30 April, 2019
- "Centre to research under-utilised crops". The Star Online. 27 June 2011. Archived from the original on 2017-09-20. Retrieved 21 August 2013
- About-crop-science: crops.org, Retrieved 5 February, 2019

Crop Genetics and Breeding

CHAPTER 2

The study of genetic variation, genes and heredity in crops is known as crop genetics. Crop breeding is the branch of science that is concerned with altering the traits of plants in order to produce desired characteristics in crops. This chapter has been carefully written to provide an easy understanding of the different techniques and technologies used within crop genetics and crop breeding.

CROP GENETICS

The science of genomics studies the entire DNA sequence of an organism, which covers all the chromosomes. Knowing the DNA sequence guides decisions about how to solve problems faced by that organism. In humans, for example, genomics is used to study diseases and seek cures. In plants, it is used to determine the best genes to be selected for improved crop varieties. The final goal is to predict which plants will perform best, and under which conditions. This allows plant breeders to select the plants most adapted to farmers' needs.

Traditional plant breeding can take years–or even decades, if you are breeding trees–to develop new crop varieties. Genomics allows scientists to identify which genes control important traits in a plant. They can determine which changes in the DNA sequence improve that trait. New technologies enable scientists to quickly study thousands of genes in thousands of plants. This allows plant breeders to choose, from a wide selection, the plants with the best DNA sequence at many genes and for many traits at the same time. This can often be done from a seed or seedling, instead of waiting for the plant to mature and show the desired leaf, flower, or fruit characteristics. This means a new variety can be developed in as few as two or three years–about half the time it used to take.

Challenges

Plant breeders are studying the genomics of plants to determine how to increase yields. They also study traits that make plants more (or less) resilient to various environmental conditions like heat and drought. Studying genomics allows scientists to look at how plants respond to disease and insect stresses. It can even help predict plants that will have higher nutritional quality. Identifying the genes, and the exact DNA sequence differences, that affect these traits can be slow and expensive. Once these DNA sequences are identified, however, selecting improved plants becomes much more efficient. Sharing information on which sequences affect which traits makes the work of all breeders and geneticists go faster.

Armyworms are a problematic pest in corn, and reduce the yield. Some corn varieties, however, are resistant to them. Finding the traits to help crops grow stronger and more reliably is an important piece of crop genetics.

Making use of advances in genomics requires that plant breeders become savvy in working with vast amounts of data. That means they need to develop better methods to store, share, and quickly analyze data. Managing the data will help produce significant advances in plant breeding.

What Advances have we Seen in Crop Genetics?

Four maize cobs, showing diversity of color and shape. Other useful features, such as resilience to pests or drought, might also be represented here. Crop genetics can help identify these qualities.

Researchers Watson and Crick discovered a way to determine the structure of DNA in 1953. In 1977, Frederick Sanger and colleagues developed a method to read the order of the base pairs, or the letters of the genetic code. Since then, ever faster methods to read the sequence and the changes in the base pairs have been developed. Genetic markers now allow the detection of single changes at known places along chromosomes. This allows scientists to know which sequences along the chromosome control which traits.

Sequencing of very long pieces of DNA, up to entire genomes of organisms, is now possible. This allows scientists to learn how each gene works together with all other genes in the organism's genome to create its final phenotype. A plant's phenotype is its appearance and how it grows and responds to changes in the environment and to other organisms. This includes insect or disease pests, beneficial insect or animal pollinators, and helpful symbionts.

Future Opportunities

Some corn is more resilient than others when it comes to diseases. Finding the traits, and passing them on to future crops, can improve food security throughout the world.

Advancing genomic technologies are making the direct editing of the DNA of crops possible for many plant breeders. Imagine a piece of ribbon with many sections. Perhaps a particular section of ribbon is worn or causing problems. It's possible to cut that part of the ribbon out and sew a new piece back in. Gene editing works like this. Scientists can determine which genes make a plant more prone to disease, or make it too tall or floppy for easy harvesting. They can 'edit' the genes and produce plants with desirable traits. It's very precise, affecting only the desired trait.

CROP BREEDING

Crop breeding is the art and science of improving important agricultural plants for the benefit of humankind. Crop breeders work to make our food, fiber, forage, and industrial crops more productive and nutritious. Crops provide for an expanding global population with increasing dietary expectations. Environmental protection is also improved by the work of crop breeders.

Plant breeding has been practiced by farmers since the dawn of agriculture, as they selected plants for larger seeds, more tasty fruits, and other valuable traits. Today, both farmers and scientists work to breed plants.

Sweeter corn, apples with a longer shelf life, and popcorn that pops better were all developed by crop breeders. But it's not just consumer preference that breeders work for.

They breed for plant growing conditions found throughout the world. Traits that have been improved by crop breeding include:

- Yield,
- Resistance to pests and diseases,
- Adaptation to environmental stresses such as heat, drought, frost, and salty soils,
- Nutritional value,
- Ease of harvest,
- Efficiency of breeding techniques,
- Taste, color, and texture,
- Creation of seedless varieties of fruits and vegetables.

Challenges

Humans around the world have an increasing appetite for diverse and nutritious foods. Feeding the world is no easy task! The answer lies in using the croplands we have in the most sustainable ways. Crop breeders work to develop crop varieties that are more productive and more nutritious, despite challenging environmental conditions. Breeding projects might take these factors into consideration:

- Plants that conserve water and soil: These precious resources are limited and in demand.
- Plants that conserve genetic diversity: The broader our genetic diversity, the more resilient our crops can be against the next disease or natural disaster.
- Plants that have better nutritional quality: More nutrition per calorie makes the best use of resources.
- Plants that produce more on the same or less land: We need to limit the further expansion of croplands to preserve our forests and other wild areas.

- Plants that are adaptable: Breeders also work to adapt our crops to rising temperatures and increasingly inconsistent weather.

These challenges are out there—and changing every day.

Future Opportunities

Crop breeding is a rapidly advancing science. It is able to make use of genetic and biotechnological innovations to efficiently develop better crop varieties.

Recent biotechnological developments are helping breeders make desired genetic changes with much greater precision. Breeders can cut out, add in, or otherwise "edit" genes so a plant can be more productive. Techniques such as marker-assisted selection mean plant breeders can quickly choose plants with desirable traits.

In addition, crop breeders gather a lot of information about the unique qualities of each plant. This means plant breeders have to be savvy in the art of working with vast amounts of data. Developing methods to store, share, and quickly analyze these data will produce significant advances in plant breeding.

SCOPE AND IMPORTANCE IN CROP IMPROVEMENT

Tissue-culture techniques are part of a large group of strategies and technologies, ranging through molecular genetics, recombinant DNA studies, genome characterization, gene-transfer techniques, aseptic growth of cells, tissues, organs and in vitro regeneration of plants that are considered to be plant biotechnologies. The use of the term biotechnology has become widespread recently but, in its most restricted sense, it refers to the molecular techniques used to modify the genetic composition of a host plant, i.e. genetic engineering.

Plant Breeding and Biotechnology

Plant breeding can be conveniently separated into two activities: manipulating genetic variability and plant evaluation. Historically, selection of plants was made by simply harvesting the seeds from those plants that performed best in the field. Controlled pollination of plants led to the realization that specific crosses could result in a new generation that performed better in the field than either of the parents or the progeny of subsequent generations, i.e. the expression of heterosis through hybrid vigour was observed. Because one of the two major activities in plant breeding is manipulating genetic variability, a key prerequisite to successful plant breeding is the availability of genetic diversity. It is in this area, creating genetic diversity and manipulating genetic variability, that biotechnology including tissue-culture techniques is having its most significant impact. In spite of the general lack of integration of most plant-biotechnology and plant-breeding programmes, field trials of transgenic plants have recently become much more common. More than 50 different plant species have already been genetically modified, either by vector-dependent (e.g. Agrobacterium) or vector-independent (e.g. biolistic, micro-injection and liposome) methods. In almost all cases, some type of tissue-culture technology has been used to recover the modified cells or tissues. In fact, tissue-culture techniques have played a major role in the development of plant genetic engineering. Tissue culture will continue to play a key role in the genetic-engineering process for the foreseeable future, especially in efficient gene transfer and transgenic plant recovery.

Wide Hybridization

A critical requirement for crop improvement is the introduction of new genetic material into the cultivated lines of interest, whether via single genes, through genetic engineering, or multiple genes, through conventional hybridization or tissue-culture techniques. During fertilization in angiosperms, pollen grains must reach the stigma of the host plant, germinate and produce a pollen tube. The pollen tube must penetrate the stigma and style and reach the ovule. The discharge of sperm within the female gametophyte triggers syngamy and the two sperm nuclei must then fuse with their respective partners. The egg nucleus and fusion nucleus then form a developing embryo and the nutritional endosperm, respectively. This process can be blocked at any number of stages, resulting in a functional barrier to hybridization and the blockage of gene transfer between the two plants.

Pre-zygotic barriers to hybridization (those occurring prior to fertilization), such as the failure of pollen to germinate or poor pollen-tube growth, may be overcome using in vitro fertilization. Post-zygotic barriers (occurring after fertilization), such as lack of endosperm development, may be overcome by embryo, ovule or pod culture. Where fertilization cannot be induced by in vitro treatments, protoplast fusion has been successful in producing the desired hybrids. In vitro fertilization IVF has been used to facilitate both interspecific and intergeneric crosses, to overcome physiological-based self incompatibility and to produce hybrids. A wide range of plant species has been recovered through IVF via pollination of pistils and self and cross-pollination of ovules. This range includes agricultural crops, such as tobacco, clover, com, rice, cole, canola, poppy and cotton. The use of delayed pollination, distant hybridization, pollination with abortive or irradiated pollen, and physical and chemical treatment of the host ovary have been used to induce haploidy.

Embryo Culture

The most common reason for post-zygotic failure of wide hybridization is embryo abortion due to poor endosperm development. Embryo culture has been successful in overcoming this major barrier as well as solving the problems of low seed set, seed dormancy, slow seed germination, inducing embryo growth in the absence of a symbiotic partner, and the production of monoploids of barley. The breeding cycle of Iris was shortened from 2 to 3 years to a few months by employing embryo rescue technology. A similar approach has worked with orchids and roses and is being applied to banana and Colocasia. Interspecific and intergeneric hybrids of a number of agriculturally important crops have been successfully produced, including cotton, barley, tomato, rice, jute, Hordeum X Secale, Triticum x Secale, Tripsacumx lea and some Brassicas. At least seven Canadian barley cultivars (Mingo, Rodeo, Craig, Winthrop, Lester and TB891-6) have been produced out of material selected from doubled haploids originating through the widely-used bulbosum method of cross-pollination and embryo rescue. Briefly, Hordeum vulgare ($2n = 14$) is pollinated with pollen from H. bulbosum ($2n = 14$). Normally, the seeds develop for about 10 days and then abort but, if the immature embryos are rescued and cultured on basal growth medium, plants can be recovered. The plants resulting from this crosspollination/embryo rescue are haploids rather than hybrids and are the result of the systematic elimination of the H. bulbosum chromosomes. Haploid wheat has also been produced by this technique.

Protoplast Fusion

Protoplast fusion has often been suggested as a means of developing unique hybrid plants which cannot be produced by conventional sexual hybridization. Protoplasts can be produced from many plants, including most crop species. However, while any two plant protoplasts can be fused by chemical or physical means, production of unique somatic hybrid plants is limited by the ability to regenerate the fused product and sterility in the interspecific hybrids rather than the production of protoplasts. Perhaps the best example of the use of protoplasts to improve crop production is that of Nicotiana, where the somatic hybrid products of a chemical fusion of protoplasts have been used to modify the alkaloid and disease-resistant traits of commercial tobacco cultivars.

Somatic hybrids were produced by fusing protoplasts, using a calcium-polyethylene glycol treatment, from a cell suspension of chlorophyll-deficient N. rustica with an albino mutant of N. tabacum. The wild N. rustica parent possessed the desirable traits of high alkaloid levels and resistance to black root rot. Fusion products were selected as bright green cell colonies, the colour being due to the genetic complemention for chlorophyll synthesis the hybrid cells. Plants recovered by shoot organogenesis showed a wide range of leaf alkaloid content but had a high level of sterility. However, after three backcross generations to the cultivated N. tabacum parent, plant fertility was restored in the hybrid lines, although their alkaloid content and resistance to blue mould and black root rot were highly variable. Interestingly, neither parent was known to possess significant resistance to blue mould.

Where mutant cell lines of donor plants are not available for use in a genetic complementation selection system, it has been demonstrated that mesophyll protoplasts from donor parents carrying transgenic antibiotic resistance can be used to produce fertile somatic hybrids selected by dual antibiotic resistance. The fusion of protoplasts from 6-azauracil-resistant cell lines of Solanum melongena (aubergine) with protoplasts from the wild species S. sisymbrilfolium yielded hybrid, purple-pigmented cell colonies that underwent regeneration via organogenesis. As protoplasts from the parental cell suspension cultures could not be regenerated, hybrids could be screened by their 6-azauracil resistance, capacity to synthesize anthocyanins (purple pigment) and ability to undergo shoot organogenesis. The restoration of regeneration ability through complementation has also been observed in Nicotiana cell-fusion products. The hybrids resulting from this study were found to be resistant to root knot nematodes and spider mites, important agricultural traits. However, they were also completely sterile and could not be incorporated into an aubergine-breeding programme. Two possible ways of solving this sterility problem, 'back' fusions of somatic hybrids with the cultivated parents and initiation of suspension cultures of the hybrid cells so that more of the wild species chromosomes can be eliminated, have so far been unsuccessful with these hybrids. Selection of hybrids and use of protoplast fusion for hybridization in crop plants has been reported in Brassicas, citrus, rice, carrot, canola, tomato, and the forage legumes alfalfa and clover. Evans & Bravo have recommended that production of novel hybrids through protoplast fusion should focus on four areas:

- Agriculturally important traits,
- Achieving combinations that can only be accomplished by protoplast fusion,
- Somatic hybrids integrated into a conventional breeding programme,
- The extension of protoplast regeneration to a wider range of crop species.

Haploids

Haploid plants are of interest to plant breeders because they allow the expression of simple recessive genetic traits or mutated recessive genes and because doubled haploids can be used immediately as homozygous breeding lines. The efficiency in producing homozygous breeding lines via doubled in vitro-produced haploids represents significant savings in both time and cost compared with other methods. Three in vitro methods have been used to generate haploids:

- Culture of excised ovaries and ovules,
- The bulbosum technique of embryo culture,
- Culture of excised anthers and pollen.

A present, 171 plant species have been used to produce haploid plants by pollen, microspore and anther culture. These include cereals (barley, maize, rice, rye, triticale and wheat), forage crops (alfalfa and clover), fruits (grape and strawberry), medicinal plants (Digitalis and Hyoscyamus), ornamentals (Gerbera and sunflower), oil seeds (canola and rape), trees (apple, litchi, poplar and rubber), plantation crops (cotton, sugar cane and tobacco), and vegetable crops (asparagus, brussels sprouts, cabbage, carrot, pepper, potato, sugar beet, sweet potato, tomato and wing bean). Haploid wheat cultivars, derived from anther culture, have been released in France and China. Five to 7 years were saved producing inbred lines in a Chinese maize-breeding programme by using anther culture-derived haploids. A similar saving has been reported for triticale and the horticultural crop Freesia. In asparagus, anther-derived haploids have been used to produce an all-male F, hybrid variety in France.

Somaclonal Variation

In addition to the variants/mutants (cell lines and plants) obtained as a result of the application of a selective agent in the presence or absence of a mutagen, many variants have been obtained through the tissue-culture cycle itself. These soma clonal variants, which are dependent on the natural variation in a population of cells, may be genetic or epigenetic, and are usually observed in the regenerated plantlets. Somaclonal variation itself does not appear to be a simple phenomenon, and may reflect pre-existing cellular genetic differences or tissue culture- induced variability. The variation may be generated through several types of nuclear chromosomal re-arrangements and losses, gene amplification or de-amplification: nonreciprocal mitotic recombination events, transposable element activation, apparent point mutations, or re-activation of silent genes in multigene families, as well as alterations in maternally inherited characteristics. Many of the changes observed in plants regenerated invitro have potential agricultural and horticultural significance. These include alterations in plant pigmentation, seed yield, plant vigour and size, leaf and flower morphology, essential oils, fruit solids and disease tolerance or resistance. Such variations have been observed in many crops, including wheat, triticale, rice, oats, maize, sugar cane, alfalfa, tobacco, tomato, potato, oilseed rape and celery. The same types of variation obtained from somatic cells and protoplasts can also be obtained from gametic tissue. One of the major potential benefits of somaclonal variation is the creation of additional genetic variability in co adapted, agronomically useful cultivars, without the need to resort to hybridization. This method could be valuable if selection is possible in vitro, or if rapid plant-screening methods are available. It is believed that somaclonal variants can be enhanced for some characters during culture in vitro, including resistance to disease pathotoxins and herbicides and tolerance to environmental or chemical stress. However, at present few cultivars of any agronomically important crop have been produced through the exploitation of somaclonal variation.

Micropropagation

During the last 30 years it has become possible to regenerate plantlets from explants and callus from all types of plants. As a result, laboratory-scale micropropagation protocols are available for a wide range of species and at present micropropagation is the widest use of plant tissue-culture technology. The cost of the labour needed to transfer tissue repeatedly between vessels and the need for asepsis can account for up to 70% of the production costs of micropropagation. Problems of vitrification, acclimatization and contamination can cause great losses in a tissue-culture laboratory. Genetic variations in cultured lines, such as polyploidy, aneuploidy and mutations, have been reported in several systems and resulted in the loss of desirable economic traits in the tissue-cultured products. There are three methods used for micropropagation:

- Enhancing axillary-bud breaking,
- Production of adventitious buds,
- Somatic embryogenesis. In the latter two methods, organized structures arise directly on the explant or indirectly from callus.

Axillary-bud breaking produces the least number of plantlets, as the number of shoots produced is controlled by the number of axillary buds cultured, but remains the most widely used method in commercial micropropagation and produces the most true to-type plantlets. Adventitious budding has a greater potential for producing plantlets, as bud primordia may be formed on any part of the inoculum. Unfortunately, somatic embryogenesis, which has the potential of producing the largest number of plantlets, can only presently be induced in a few species.

Synthetic Seed

A synthetic or artificial seed has been defined as a somatic embryo encapsulated inside a coating and is considered to be analogous to a zygotic seed. There are several different types of synthetic seed: somatic embryos encapsulated in a water gel; dried and coated somatic embryos; dried and uncoated somatic embryos; somatic embryos suspended in a fluid carrier; and shoot buds encapsulated in a water gel. The use of synthetic seeds as an improvement on more traditional micropropagation protocols in vegetatively propagated crops may, in the long term, have tissue culture and crop improvement a cost saving, as the labour intensive step of transferring plants from in vitro to soil/field conditions may be overcome. Other applications include the maintenance of male sterile lines, the maintenance of parental lines for hybrid crop production, and the preservation and multiplication of elite genotypes of woody plants that have long juvenile developmental phases. However, before the widespread application of this technology, somaclonal variation will have to be minimized, large-scale production of high quality embryos must be perfected in the species of interest, and the protocols will have to be made cost-effective compared with existing seed or micropropagation technologies.

Germplasm Preservation

One way of conserving germplasm, an alternative to seed banks and especially to field collections of clonally propagated crops, is in vitro storage under slow-growth conditions (at low temperature and/or with growth-retarding compounds in the medium) or cryopreservation or as desiccated synthetic seed. The technologies are all directed towards reducing or stopping growth and metabolic activity. Techniques have been developed for a wide range of plants. The most serious limitations are a lack of a common method suitable for all species and genotypes, the high costs and the possibility of somaclonal variation and non-intentional cell-type selection in the stored material (e.g. aneuploidy due to cell division at low temperatures or non-optimal conditions giving one cell type a selective growth advantage.

Plant tissue-culture technology is playing an increasingly important role in basic and applied studies, including crop improvement. In modern agriculture, only about 150 plant species are extensively cultivated. Many of these are reaching the limits of their improvement by traditional methods. The application of tissue-culture technology, as a central tool or as an adjunct to other methods, including recombinant DNA techniques, is at the vanguard in plant modification and improvement for agriculture, horticulture and forestry.

TECHNOLOGIES TO BOOST BREEDING

Breeding crops with a high yield and superior adaptability is vital to maintaining global food security. New technologies on multiple scales are re-engineering traditional plant breeding to meet these challenges.

Although practised for thousands of years, crop breeding remains a demanding and time-consuming task. A major problem is the long generation time of crops, as breeding needs frequent crossing and selfing. The technical rationale is to shorten the generation cycle by extending the photoperiod using supplemental LED lighting. This represents an intriguing step forwards, but is only one of many technological breakthroughs that promise to boost modern breeding.

Breeding is nothing more nor less than directed evolution under artificial selection. Significant progress has been made in conventional breeding methods by utilizing desirable genetic variations. Semi-dwarfness was bred into modern rice and wheat varieties, leading to high-yield and lodging-resistant varieties; maize was biofortified with increased beta-carotene; the non-ripening tomato was developed with extended shelf-life.

Most agronomic traits have a complex genetic nature; what may seem to be simple traits rely on the combination of multiple genes. Worse still, most genes display pleiotropic effects so improving one trait can adversely affect others. A breeder's main task is labour-intensively seeking the optimal combination of genes for particular conditions.

The development of molecular and quantitative genetics offers a good opportunity to accelerate breeding by selecting either for known genes or quantitative trait loci using genetic markers, rather than effects on phenotypes. Last year Zeng et al. reported the development of super rice varieties based on known elite alleles that are associated with grain quality and yield by crossing three parental lines in five years. This was a demonstration of the development of elite crop varieties though well-designed marker-assisted selection (MAS), the availability of high-throughput genotyping making the MAS simpler to implement.

Another route is to harness heterosis. Hybrid breeding revolutionized agriculture in the twentieth century. In 1974 the first hybrid rice variety was released in China, leading to a yield gain of 20–30%. The use of heterosis in maize breeding is even older. In the early 1900s, maize hybrids were already commercialized due to their superior performance. No single gene or subset of genes can explain heterosis in all systems. different rice hybrid breeding systems used different loci.

Despite the success of conventional and hybrid breeding, yield increases may have plateaued for staple crops. This is surprising given that scientists have uncovered many genes that are claimed to increase yield. Desirable alleles may be lacking or existing gene pools underexplored; alleles found by functional studies to affect phenotypes may be disappointing when bred into commercialized varieties; polyploid crops can be resistant to breeding gain and efficient methods of selection may be lacking. Looking forwards, genome editing and genomic selection may boost breeding.

Traditional mutagenesis methods produce random mutations across genomes, but these mutant alleles are rarely agronomically meaningful. Etienne Bucher and colleagues recently proposed a method for stress-controlled mobilization of retrotransposons, representing a semi-targeted approach for generating new mutations. Targeted mutagenesis technologies are now widely used in plants, especially the CRISPR technologies, which can generate desirable mutations in a swift and convenient manner. For example, Caixia Gao and Jinlong Qiu showed that genome editing can engineer three homoalleles simultaneously in hexaploid wheat to confer resistance to powdery mildew. By using single guide RNAs (sgRNAs) targeting virus DNA, CRISPR can also engineer plant resistance against viruses.

For more complex traits, if the regulatory genes or sequences have been characterized CRISPR can be used in combination with genetic design to generate a series of alleles with a continuous range of phenotypes for screening, as demonstrated recently by Zachary Lippmann's group8. They showed rapid generation of cis-regulatory alleles in tomato that confer a continuum of phenotypic variation for yield-related traits. The growing toolbox of genome editing will allow the generation of more variations, providing even greater potential for producing desirable agronomic traits.

Genomic selection (GS), is theoretically similar to MAS but it uses high-density genome-wide markers. GS employs predictive computational models that are developed from a training population to make genome-wide predictions of phenotypes in a breeding population. GS examples often use biparental or multiparental populations.

Another typical application of GS is to turbocharge underexploited genebanks. In 2016, Jianming Yu's group showed that high-accuracy models can be developed to predict the traits of sorghum accessions in gene banks, demonstrating a strategy to assess and utilize valuable but poorly explored germplasms. The same year saw genomic predictions of wheat landraces, and soybean accessions although with comparatively lower accuracies. High-throughput phenotyping platforms provide an opportunity to develop accurate prediction models at a lower cost.

CONVENTIONAL PLANT BREEDING

Since the practice of agriculture began, eight to ten thousand years ago, farmers have been altering the genetic makeup of the crops they grow. Early farmers selected the best looking plants and seeds and saved them to plant for the next season. Then, once the science of genetics became better understood, plant breeders used what they knew about the genes of a plant to select for specific desirable traits to develop improved varieties.

The selection for features such as faster growth, higher yields, pest and disease resistance, larger seeds, or sweeter fruits has dramatically changed domesticated plant species compared to their wild relatives. For example, when corn was first grown in North and South America, thousands of years ago, the corn cobs farmers harvested were smaller than one's little finger. Today, there are hundreds of varieties of corn, some of which produce cobs as long as one's forearm.

Conventional plant breeding has been going on for hundreds of years, and is still commonly used today. Early farmers discovered that some crop plants could be artificially mated or cross-pollinated to increase yields. Desirable characteristics from different parent plants could also be combined in the offspring. When the science of plant breeding was further developed in the 20th century, plant breeders understood better how to select superior plants and breed them to create new and improved varieties of different crops. This has dramatically increased the productivity and quality of the plants we grow for food, feed and fiber.

The art of recognizing desirable traits and incorporating them into future generations is very important in plant breeding. Breeders scrutinize their fields and travel long distances in search of individual plants that exhibit desirable traits. A few of these traits occasionally arise spontaneously through a process called mutation, but the natural rate of mutation is very slow and unreliable to produce all the plant traits that breeders would like to see.

Hybrid Seed Technology

The end result of plant breeding is either an open-pollinated (OP) variety or an F1 (first filial generation) hybrid variety. OP varieties, when maintained and produced properly, retain the same characteristics when multiplied. The only technique used with OP varieties is the selection of the seed-bearing plants.

Hybrid seeds are an improvement over open pollinated seeds in terms of qualities such as yield, resistance to pests and diseases, and time to maturity. Hybrid seeds are developed by the hybridization or crossing of parent lines that are 'pure lines' produced through inbreeding. Pure lines are plants that "breed true" or produce sexual offspring that closely resemble their parents. By crossing pure lines, a uniform population of F1 hybrid seed can be produced with predictable characteristics.

The simplest way to explain how to develop an F1 hybrid is to take an example. Let us say a plant breeder observes a particularly good habit in a plant, but with poor flower color, and in another plant of the same type he sees good color but poor habit. The best plant of each type is then taken and self-pollinated (in isolation) each year and, each year, the seed is re-sown. Eventually, every time the seed is sown the same identical plants will appear. When they do, this is known as a 'pure line.'

If the breeder now takes the pure line of each of the two plants he originally selected and cross pollinates the two by hand the result is known as an "F1 hybrid." Plants are grown from the seed produced, and the result of this cross pollination should have the combined traits of the two parents.

This is the simplest form of hybridization, but there are complications, of course. A completely pure line can sometimes take seven or eight years to achieve. Sometimes, a pure line is made up of several previous crossings to build in desirable features. The resulting plant is then grown on until it is genetically pure before use in hybridization.

In addition to qualities like good vigor, trueness to type, heavy yields and high uniformity which hybrid plants enjoy, other characteristics such as earliness, disease and insect resistance and good water holding ability have been incorporated into most F1 hybrids.

Unfortunately, these advantages come with a price. Because creating F1 hybrids involves many years of preparation to create pure lines that have to be constantly maintained so that F1 seeds can be harvested each year, the seeds then become more expensive. The problem is compounded because to ensure that no self-pollination takes place, all the hybridization of the two pure lines, sometimes, has to be done by hand.

Another disadvantage is if the seeds of the F1 hybrids are used for growing the next crops, the resulting plants do not perform as well as the F1 material - resulting in inferior yields and vigor. As a consequence, the farmer has to purchase new F1 seeds from the plant breeder each year. The farmer is, however, compensated by higher yields and better quality of the crop.

Though more expensive, hybrid seeds have had a tremendous impact on agricultural productivity. Today, nearly all corn and 50% of all rice are hybrids (DANIDA).

In the US, the widespread use of corn hybrids, coupled with improved cultural practices by farmers, has more than tripled corn grain yields over the past 50 years from an average of 35 bushels per acre in the 1930s to 115 bushels per acre in the 1990s. No other major crop anywhere in the world even comes close to equaling that sort of success story.

Hybrid rice technology helped China increase its rice production from 140 million tons in 1978 to 188 million tons in 1990. hybrid rice technology offers opportunities for increasing rice varietal yields by 15-20%. And this is achievable with the improved, semi-dwarf, and inbred varieties.

Many cultivars of popular vegetables or ornamental plants are F1 hybrids. In terms of improved plant characteristics, tropical vegetable breeders can point to some rather clear achievements over the last two decades:

- Yield improvement: Hybrids often outyield traditional OP selections by 50-100% due to its improved vigor, improved genetic disease resistance, improved fruit setting under stress, and higher female/male flower ratios.
- Extended growing season: Hybrids often mature up to 15 days earlier than local OP varieties. For many crops, the hybrid's relative advantage over the OP is most pronounced under stress conditions.
- Quality improvement: Hybrids have helped stabilize product quality at a higher, and more uniform level – this implies improved consumption quality (e.g. firm flesh of wax gourd, crispy taste of watermelon).

BACKCROSSING

Backcrossing is a crossing of a hybrid with one of its parents or an individual genetically similar to its parent, in order to achieve offspring with a genetic identity which is closer to that of the parent. It is used in horticulture, animal breeding and in production of gene knockout organisms.

Backcrossed hybrids are sometimes described with acronym "BC", for example, an F1 hybrid crossed with one of its parents (or a genetically similar individual) can be termed a BC1 hybrid, and a further cross of the BC1 hybrid to the same parent (or a genetically similar individual) produces a BC2 hybrid.

Advantages

- If the recurrent parent is an elite genotype, at the end of the backcrossing programme an elite genotype is recovered.
- As there is no "new" recombination, the elite combination is not lost.

Disadvantages

- Works poorly for quantitative traits,
- Is more restricted for recessive traits,
- In practice, sections of genome from the non-recurrent parents are often still present and can have unwanted traits associated with them,
- For very wide crosses, limited recombination may maintain thousands of 'alien' genes within the elite cultivar,
- Many backcrosses are required to produce a new cultivar which can take many years.

Natural Backcrossings

York radiate groundsel (*Senecio eboracensis*) is a naturally occurring hybrid species of Oxford ragwort (*Senecio squalidus*) and common groundsel (*Senecio vulgaris*). It is thought to have arisen from a backcrossing of the F1 hybrid with *S. vulgaris*.

Again, the pure tall (TT) and pure dwarf (tt) pea plants when crossed in the parental generation, they produce all heterozygote (Tt) tall pea plants in the first filial generation. The cross between first filial heterozygote tall (Tt) pea plant and pure tall (TT) or pure dwarf (tt) pea plant of the parental generation is also an example for the back-crossing between two plants. In this case, the filial generation formed after the back cross may have a phenotype ratio of 1:1 if the cross is made with recessive parent or else all offspring may be having phenotype of dominant trait if back cross is with parent having dominant trait. The former of these traits is also called a test cross.

Artificially Recombinant Lines

In plants, *inbred backcross lines* (IBLs) refers to lines (i.e. populations) of plants derived from the repeated backcrossing of a line with artificially recombinant DNA with the wild type, operating some kind of selection that can be phenotypical or through a molecular marker (for the production of introgression lines).

MOLECULAR BREEDING AND MARKER-ASSISTED SELECTION

The process of developing new crop varieties can take almost 25 years. Now, however, biotechnology has considerably shortened the time to 7-10 years for new crop varieties to be brought to the market. One of the tools which can make it easier and faster for scientists to select plant traits is marker-assisted selection (MAS).

Molecular Shortcut

The differences that distinguish one plant from another are encoded in the plant's genetic material, the DNA. DNA is packaged in chromosome pairs (strands of genetic material), one coming from each parent. The genes, which control a plant's characteristics, are located on specific segments of each chromosome. Together, all of a plant's genes make up its genome.

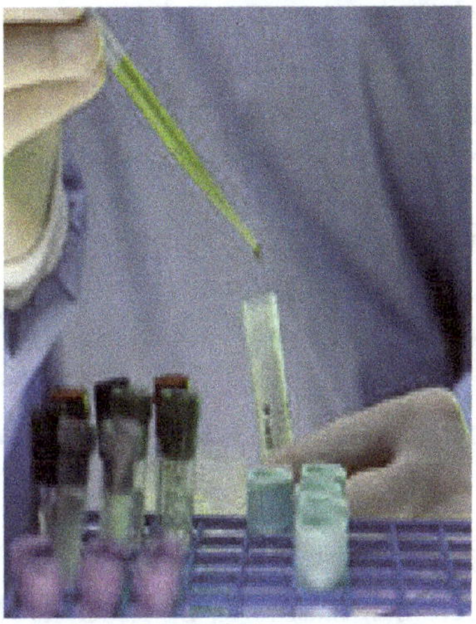

Some traits, like flower color, may be controlled by only one gene. Other more complex characteristics, however, like crop yield or starch content, may be influenced by many genes. Traditionally, plant breeders have selected plants based on their visible or measurable traits, called the phenotype. This process can be difficult, slow, influenced by the environment, and costly – not only in the development itself, but also for the economy, as farmers suffer crop losses.

As a shortcut, plant breeders now use marker-assisted selection (MAS). To help identify specific genes, scientists use what are called molecular or genetic markers. The markers are a string or sequence of nucleic acid which makes up a segment of DNA. The markers are located near the DNA sequence of the desired gene and are transmitted by the standard laws of inheritance from one generation to the next. Since the markers and the genes are close together on the same chromosome, they tend to stay together as each generation of plants is produced. This is called genetic linkage. This linkage helps scientists to predict whether a plant will have a desired gene. If researchers can find the marker for the gene, it means the desired gene itself is present.

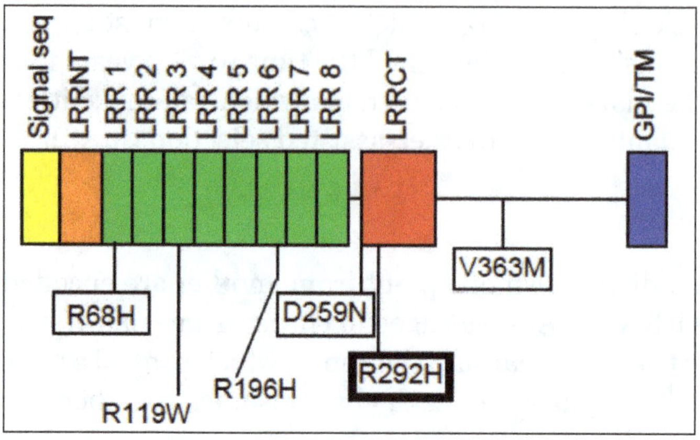

As scientists learn where markers occur on a chromosome, and how close they are to specific genes, they can create a genetic linkage map. Such a map would show the location of markers and genes, and their distance from other known genes. Scientists can produce detailed maps in only one generation of plant breeding.

Using very detailed genetic maps and better knowledge of the molecular structure of a plant's DNA, researchers can analyze only a tiny bit of plant tissue, even from a newly germinated seedling. Once the tissue is analyzed, scientists know whether that seedling contains the appropriate gene. If it doesn't, they can quickly move on and concentrate on analysis of another seedling, eventually working only with the plants which contain a specific trait.

It should be noted, however, that molecular breeding through MAS is somewhat limited in scope compared to genetic engineering or modification because: 1) it works only for traits already present in a crop; 2) it cannot be used effectively to breed crops which have long generation times (e.g. citrus); and 3) it cannot be used effectively with crops which are clonally propagated because they are sterile or do not breed true (this includes many staples such as yams, bananas, plantain, sweet potato, and cassava).

Molecular Markers

Several marker systems have been developed and are applied to a range of crop species. These are the Restriction Fragment Length Polymorphisms (RFLPs), Random Amplification of Polymorphic DNAs (RAPDs), Sequence Tagged Sites (STS), Amplified Fragment Length Polymorphisms (AFLPs), Simple Sequence Repeats (SSRs) or microsatellites, and Single Nucleotide Polymorphism (SNPs). The advantages and disadvantages of these marker systems are provided in table.

Table: Comparison of most commonly used marker systems

Feature	RFLPs	RAPDs	AFLPs	SSRs	SNPs
DNA required (?g)	10	0.02	0.5-1.0	0.05	0.05
DNA quality	High	High	Moderate	Moderate	High
PCR-based	No	Yes	Yes	Yes	Yes
Number of polymorph loci analyzed	1.0-3.0	1.5-50	20-100	1.0-3.0	1.0
Ease of use	Not easy	Easy	Easy	Easy	Easy
Amenable to automation	Low	Moderate	Moderate	High	High
Reproducibility	High	Unreliable	High	High	High
Development cost	Low	Low	Moderate	High	High
Cost per analysis	High	Low	Moderate	Low	Low

These molecular techniques have been widely used to monitor differences in DNA sequence in and among species. They also allow the creation of new sources of genetic variation by introducing new and desirable traits from wild varieties into elite lines. While RFLP markers have been the basis for most genetic work in crop plants, AFLPs and SSRs are currently the most popular techniques used due to ease in detection and automation. The adoption of the new marker system, SNPs, is now highly preferred, with the increasing amount of sequence information, and the determination of gene function due to genomic research.

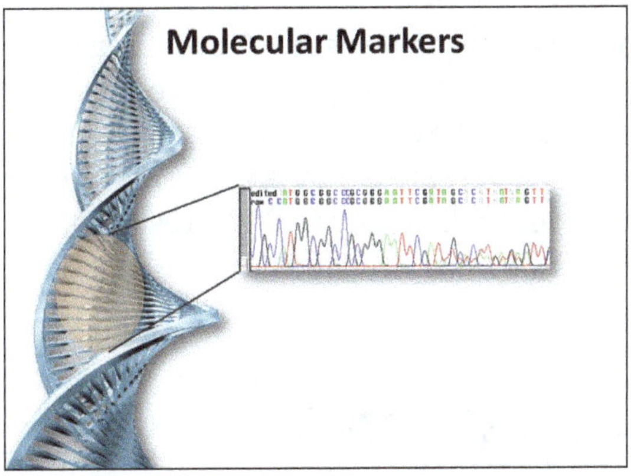

The main uses of these molecular markers in crop genetic studies are as follows:

- Assessment of genetic variability and characterization of germplasm,
- Identification and fingerprinting of genotypes,
- Estimation of genetic distances between population, inbreeds, and breeding materials,
- Detection of monogenic and quantitative trait loci (QTL),
- Marker-assisted selection,
- Identification of sequences of useful candidate genes.

MAS for Pathogen Resistance in Tomato

One of the major constraints in tomato cultivation and production are severe harvest losses caused by a number of pathogens, including viruses, bacteria, fungi, and nematodes. Farmers have adopted control measures, such as applications of agrochemicals and use of resistant lines. Although conventional breeding has had a significant impact on improving resistance of tomato, the time-consuming process of making crosses and backcrosses, and the selection of the desired resistant progeny make it difficult to react adequately to the evolution of new virulent pathogens.

Molecular markers are now being widely used for breeding tomato. More than 40 genes that confer resistance to major classes of tomato pathogens have been mapped, cloned, and/or sequenced. These maps have allowed for "pyramiding" resistance genes in tomato through MAS, where several resistance genes can be engineered into one genotype. Currently, tomato breeding through MAS has resulted in varieties with resistance or tolerance to one or more specific pathogens.

MUTATION BREEDING

Mutation breeding, sometimes referred to as "variation breeding", is the process of exposing seeds to chemicals or radiation in order to generate mutants with desirable traits to be bred with other cultivars. Plants created using mutagenesis are sometimes called mutagenic plants or mutagenic seeds. From 1930 to 2014 more than 3200 mutagenic plant varieties were released that have been derived either as direct mutants (70%) or from their progeny (30%). Crop plants account for 75% of released mutagenic species with the remaining 25% ornamentals or decorative plants. However, although the FAO/IAEA reported in 2014 that over 1,000 mutant varieties of major staple crops were being grown worldwide, it is unclear how many of these varieties are currently used in agriculture or horticulture around the world, as these seeds are not always identified or labeled as having a mutagenic provenance.

Process

There are different kinds of mutagenic breeding such as using chemical mutagens like ethyl methanesulfonate and dimethyl sulfate, radiation or transposons to generate mutants. Mutation breeding is commonly used to produce traits in crops such as larger seeds, new colors, or sweeter fruits, that either cannot be found in nature or have been lost during evolution.

Radiation Breeding

Exposing plants to radiation is sometimes called radiation breeding and is a sub class of mutagenic breeding. Radiation breeding was discovered in the 1920s when Lewis Stadler of the University of Missouri used X-rays on maize and barley. In the case of barley, the resulting plants were white, yellow, pale yellow and some had white stripes. In 1928, Stadler first published his findings on radiation-induced mutagenesis in plants. During the period 1930–2004, radiation-induced mutant varieties were developed primarily using gamma rays (64%) and X-rays (22%).

Radiation breeding may take place in atomic gardens; and seeds have been sent into orbit in order to expose them to more cosmic radiation.

Use of Chemical Mutagens

High rates of chromosome aberrations resulting from ionizing radiation and the accompanied detrimental effects made researchers look for alternate sources for inducing mutations. As a result, an array of chemical mutagens has been discovered. The most widely used chemical mutagens are alkylating agents. Ethyl methanesulfonate (EMS) is the most popular because of its effectiveness and ease of handling, especially its detoxification through hydrolysis for disposal. Nitroso compounds are the other alkylating agents widely used, but they are light-sensitive and more precautions need to be taken because of their higher volatility. EMS has become a commonly used mutagen for developing large numbers of mutants for screening such as in developing tilling populations. Although many chemicals are mutagens, only few have been used in practical breeding as the doses need to be optimised and also because the effectiveness is not high in plants for many.

Comparison to other Agronomic Techniques

In the debate over genetically modified foods, the use of transgenic processes is often compared and contrasted with mutagenic processes. While the abundance and variation of transgenic organisms in human food systems, and their effect on agricultural biodiversity, ecosystem health and human health is somewhat well documented, mutagenic plants and their role on human food systems is less well known, with one journalist writing “Though poorly known, radiation breeding has produced thousands of useful mutants and a sizable fraction of the world's crops including varieties of rice, wheat, barley, pears, peas, cotton, peppermint, sunflowers, peanuts, grapefruit, sesame, bananas, cassava and sorghum." In Canada crops generated by mutation breeding face the same regulations and testing as crops obtained by genetic engineering. Mutagenic varieties tend to be made freely available for plant breeding, in contrast to many commercial plant varieties or germplasm that increasingly have restrictions on their use such as terms of use, patents and proposed genetic user restriction technologies and other intellectual property regimes and modes of enforcement.

Unlike genetically modified crops, which typically involve the insertion of one or two target genes, plants developed via mutagenic processes with random, multiple and unspecific genetic changes have been discussed as a concern but are not prohibited by any nation's organic standards. Reports from the US National Academy of Sciences state that there is no scientific justification for regulating genetic engineered crops while not doing so for mutation breeding crops.

Several organic food and seed companies promote and sell certified organic products that were developed using both chemical and nuclear mutagenesis. Several certified organic brands, whose companies support strict labeling or outright bans on GMO-crops, market their use of branded wheat and other varietal strains which were derived from mutagenic processes without any reference to this genetic manipulation. These organic products range from mutagenic barley and wheat ingredient used in organic beers to mutagenic varieties of grapefruits sold directly to consumers as organic.

New Mutagen Techniques

Restriction Endonucleases

Interest in the use of bacterial restriction endonucleases (RE) to study double-stranded breaks in plant DNA began in the mid-nineties. These breaks in DNA, otherwise known as DSBs, were found to be the source of much chromosomal damage in eukaryotes, causing mutations in plant varieties. REs induce a result on plant DNA similar to that of ionizing radiation or radiomimetic chemicals. Blunt ended breaks in the DNA, unlike sticky ended breaks, were found to produce more variations in chromosomal damage, making them the more useful type of break for mutation breeding. While the connection of REs to chromosomal aberrations is mostly limited to research on mammalian DNA, success in mammalian studies caused scientists to conduct more studies of RE-induced chromosomal and DNA damaged on barley genomes. Due to restriction endonucleases' ability to facilitate damage in chromosomes and DNA, REs have the capability of being used as a new method of mutagenesis to promote the proliferation of mutated plant varieties.

Space-Breeding

The ability of plants to develop and thrive is dependent on conditions such as microgravity and cosmic radiation in space. China has been experimenting with this theory by sending seeds into space, testing to see if space flights will cause genetic mutations. Since 1987, China has cultivated 66 mutant varieties from space through their space-breeding program. Chromosomal aberrations greatly increased when seeds were sent into aerospace compared to their earth-bound counterparts. The effect of space flight on seeds depends on their species and variety. For example, space-bred wheat saw a large growth in seed germination in compared to its Earth-bound control, but space-bred rice had no visible advantage compared to its control. For the varieties that were positively mutated by space flight, their growth potential exceeded that of not only their Earth-grown counterparts, but also their irradiated counterparts on Earth. Compared to traditional mutagenic techniques, space-bred mutations have greater efficacy in that they experience positive effects on their first generation of mutation, whereas irradiated crops often see no advantageous mutations in their first generations. Though multiple experiments have shown the positive effects of space flight on seed mutation, there is no clear connection as to what aspect of aerospace has produced such advantageous mutations. There is much speculation around cosmic radiation being the source of chromosomal aberrations, but so far, there has been no concrete evidence of such connection. Though China's space-breeding program has been shown to be very successful, the program requires a large budget and technological support that many other countries are either unwilling or unable to provide, meaning this program is unfeasible outside of China. Due to such restraints, scientists have been trying to replicate space condition on Earth in order to promote the same expedient space-born mutations on Earth. One such replication is a magnetic field-free space (MF), which produces an area with a weaker magnetic field than that of Earth. MF treatment produced mutagenic results, and has been used to cultivate new mutant varieties of rice and alfalfa. Other replications of space conditions include irradiation of seeds by a heavy 7 Li-ion beam or mixed high-energy particles. These space-bred varieties are already being introduced to the public. In 2011, during the National Lotus Flowers Exhibition in China, a mutant lotus, called the "Outer Space Sun", was shown at the flower show.

Ion Beam Technology

Ion beams mutate DNA by deleting multiple bases from its code. Compared to traditional sources of radiation, like gamma rays and X-rays, ion beams have been shown to cause more severe breaks in DNA that are more difficult to weave back together, causing the change in DNA to be more drastic than changes caused by traditional irradiation. Ion beams change DNA in a manner that makes it look vastly different than its original makeup, more so than when traditional irradiation techniques are used. Most experimentation, using ion beam technology, has been conducted in Japan. Notable facilities using this technology are TIARA of the Japan Atomic Energy Agency,

RIKEN Accelerator Research Facility, and various other Japanese institutions. During the process of ion beam radiation, seeds are wedged between two kapton films and irradiated for roughly two minutes. Mutation frequencies are notably higher for ion beam radiation compared to electron radiation, and the mutation spectrum is broader for ion beam radiation compared to gamma ray radiation. The broader mutation spectrum was revealed through the largely varied amount of flower phenotypes produced by ion beams. Flowers mutated by the ion beams exhibited a variety of colors, patterns, and shapes. Through ion beam radiation, new varieties of plants have been cultivated. These plants had the characteristics of being ultraviolet light-B resistant, disease resistant, and chlorophyll-deficient. Ion beam technology has been used in the discovery of new genes responsible for the creation of more robust plants, but its most prevalent use is commercially for producing new flower phenotypes, like striped chrysanthemums.

Mature Pollen Treated with Gamma Radiation

Gamma radiation is used on mature rice pollen to produce parent plants used for crossing. The mutated traits in the parent plants are able to be inherited by their offspring plants. Because rice pollen has a very short lifespan, researchers had to blast gamma rays at cultured spikes from rice plants. Through experimentation, it was revealed that there was a greater variety of mutation in irradiated pollen rather than irradiated dry seeds. Pollen treated with 46Gy of gamma radiation showed an increase in grain size overall and other useful variations. Typically, the length of each grain was longer after the crossing of irradiated parent rice plants. The rice progeny also exhibited a less chalky visage, improving on the appearance of the parent rice plants. This technique was used to develop two new rice cultivars, Jiaohezaozhan and Jiafuzhan, in China. Along with facilitating the creation of these two rice cultivars, the irradiation of mature rice pollen has produced roughly two hundred mutant rice lines. Each of these lines produce rice grains of both a higher quality and larger size. The mutations produced by this technique vary with each generation, meaning further breeding of these mutated plants could produce new mutations. Traditionally, gamma radiation is used on solely adult plants, and not on pollen. The irradiation of mature pollen allows mutant plants to grow without being in direct contact with gamma radiation. This discovery is in contrast to what was previously believed about gamma radiation: that it could only elicit mutations in plants and not pollen.

INBREEDING

Inbreeding can occur in plants via selfing, or the union of haploid gametes produced by the same genetic individual, also known as autogamy. As most flowers are hermaphroditic, selfing is a reproductive strategy present in the majority of plants. Furthermore, many plants have multiple flowers or inflorescences per individual, allowing for another type of self-fertilization, termed geitonogamy. Darwin showed in early studies that this type of selfing differs from selfing within flowers in terms of the quality of offspring produced. Because plants can reproduce vegetatively and apomictically (seeds having the same genotype as their mother) discrete individuals can actually be genetic clones. Therefore, another class of selfing will be mating between genetically identical individuals. Inbreeding of course can also occur in plants through the mating between close relatives, which is termed biparental inbreeding to differentiate it from selfing. Where dispersal of propagules is limited, genetically related individuals tend to occur in patches or clumps across a landscape, increasing biparental inbreeding. Similarly, when small, less vagile insects achieve pollen dispersal, pollination between relatives within a patch will prevail.

Inbreeding Depression

With some exceptions, inbreeding reduces offspring fitness in essentially all naturally outcrossing plants and to a lesser extent in selfing species. The negative effects of mating between relatives have been noticed for many centuries. The careful breeding studies of Darwin first empirically demonstrated inbreeding depression in a wide variety of taxa. The negative effects of inbreeding have since been observed in both outcrossing and selfing species for a variety of traits with consequences for offspring fitness. Examples of traits shown to be subjects to inbreeding depression include pollen quantity, number of ovules, amount of seed, germination rate, growth rate and competitive ability. Genetic models have been developed to functionally explain reduction in fitness-related traits caused by inbreeding.

The primary genetic consequence of inbreeding is increased homozygosity. Two hypotheses for the genetic basis of inbreeding depression are put forth, both of which depend on the idea that homozygosity wll increase with inbreeding. Either the overdominance or partial dominance hypotheses are invoked to model the negative consequences of inbreeding. In the overdominance hypothesis, inbreeding depression is attributable to higher fitness of heterozygotes versus homozygotes for the loci in question. For the partial dominance hypothesis, negative fitness consequences for inbred lines are due to the fixation of recessive or partially recessive deleterious alleles. Current thought favors the latter hypothesis, where inbreeding depression is attributable to many genes of small effect. However, distinguishing between the two genetic mechanisms is complicated by linked sets of deleterious recessives that imitate overdominance effects.

The Evolution of Inbreeding in Plants

Because self-fertilization will produce the most profound inbreeding, discussion of the evolution of inbreeding in plants primarily focuses on the evolution of selfing. Altogether, plants provide an excellent system in which to study the evolution of inbreeding. Members of the same genus that are otherwise closely related often display a variety of breeding systems. The ability to self-fertilize is almost universally a derived characteristic, with self-fertilizing species arising from outcrossing species. By comparing predominantly outcrossing versus selfing species, the selective pressure surrounding the evolution of self-fertilization versus outcrossing can be estimated.

The evolution of inbreeding in plants is the product of the interplay between several selective forces, with inbreeding depression being the primary cost associated with self-fertilization. Outcrossing has inherent costs for the parent plant that will influence fitness. For example, the direct costs of copious pollen production for effective fertilization take away from the potential for seed production. The so-called 'meiotic cost' for sexual reproduction occurs because only transmitting half of the genome to the next generation lowers parental fitness. Outcrossing will also have fitness costs for offspring if it breaks up co-adapted gene complexes that confer adaptation to a local environmental conditions, i.e. outbreeding depression.

Various other ecological phenomena also foster the evolution of selfing in plants. For example, stochastic fluctuations in pollinator populations could lead to reduced seed set for insect cross-pollinated plants in years with low pollinator numbers. Populations that are subject to environmental stochasticity benefit by selfing when years are either too dry or too wet for pollen to be spread between plants. Also, founding of new populations by single individual, as is evidenced in many ruderal (weedy) species, favors self-fertilizing species. In the absence of inbreeding depression via partial dominance, selection against outcrossing would lead to the predominance of selfing. However, outcrossing is necessary to maintain genetic variability to ensure the long-term evolutionary potential of a species.

Quantitative genetic models have been developed for the evolution of inbreeding and self-fertilization in plants. Fisher first demonstrated that a gene leading to self-fertilization would rapidly spread in an outcrossing population until the entire population was selfing. However, it is unlikely that selfing is conferred by a single gene of major effect, but rather a shift to selfing is likely due to a suite of genes of small effect. As demonstrated by Lande and Schemske predominant outcrossing and predominant selfing can exist as alternative stable states of a mating system. Their quantitative genetic models also confirm that highly self-fertilizing species can evolve more readily from primarily outcrossing species than vice versa. However, primarily selfing species are not necessarily an evolutionary dead end as sufficient quantitative variation can be maintained between inbred lines of a species.

How Inbreeding Depression is Measured in Plants

Breeding Studies

In selfing plant populations, the ratio of the fitness of selfed versus outcrossed progeny is used as the measure of inbreeding depression. In Darwin's classic studies of the effects of inbreeding, carefully controlled crosses were used to determine the effects of inbreeding. Controlled crosses remain to this day the primary method for measuring inbreeding depression. Generally, crossing individuals of a known or inferred relatedness produces outcrossed offspring in these types of studies. Where the fine-scale genetic structure or pedigree of a natural population is unknown, distance between parental plants is used as a proxy measure of relatedness. The inbred offspring can be similarly produced by crossing individuals within a short distance of one another, or by enclosing hermaphroditic flowers to allow them to self while preventing outcrossing. Differences in seed set between outcrossed and inbred flowers can indicate inbreeding depression if fewer seeds are produced through inbreeding. In plants without gametophytic self-incompatibility loci, reduced seed set in inbred flowers in is often attributable to lethal recessives that prevent embryo formation or other essential processes for seed set.

Measuring Inbreeding in Natural Populations

The breeding studies described above are amenable to greenhouse experiments, where manipulation of pollination can be easily achieved and controlled. However, a study by Dudash indicates that offspring fitness measured in greenhouse or garden environments may significantly underestimate inbreeding depression. More indirect methods are needed to study the effects of inbreeding depression in natural populations. Inbreeding depression has been investigated at the population level in several recent studies by examining how fitness-related traits vary between small and large populations. Reduced fitness in populations may not always be attributable to inbreeding depression due to confounding environmental variation. Experimental studies have also demonstrated density-dependent seed set known as the Allee effect to occur in small populations of plants.

Molecular Methods

Studies have made use of advances in molecular genetics for detecting inbreeding in plant populations. The advantage of this approach is that pedigree information and complex breeding experiments are not always requisite, which are often difficult to apply to natural populations. The similarity in state between two alleles at a locus determined by molecular markers can be used to estimate F, the inbreeding coefficient. Changes in F across two generations of a selfing plant can be used to estimate selection against inbred individuals, i.e. inbreeding depression. Heterozygosity at marker loci is assumed to be indicative of general levels of heterozygosity in the genome, which is expected to decrease with inbreeding . Positive correlations between fitness-related traits and heterozygosity are typically taken as evidence for inbreeding depression. The application of quantitative trait loci (QTL) methods to the issue of inbreeding depression is in its early stages of development, but has provided significant clues to its underlying genetic basis QTLs allow for the association of observable inbreeding depression effects with discrete regions of the genome. For example in Pinus taeda, genes of major effect detected by marker linkage were primarily responsible for inbreeding depression; this result supports the partial dominance hypothesis.

Inbreeding Effects in Wild Populations

The effects of inbreeding in natural population are of interest to ecologists seeking to determine the underlying causes for observed demographic phenomena. Population level investigations of inbreeding have recently documented these effects in a number of species. Menges compared populations of an endemic prairie species Silene regia (royal catchfly) and found that seed set increased with population size. In a related study, Oostermeijer and colleagues found that offspring fitness of Gentiana pneumonanthe (marsh gentian) increased with population size. In a related rare species, Gentianella germanica (chiltern gentian), steeper population declines via reduced seed set were found in smaller populations; this reduced plant fitness occurred independently of environmental variation, indicating the inbreeding effects were the primary cause. As indicated in a study of Ipomopsis agregata (scarlet gilia), the effects of inbreeding depression may make populations more susceptible to environmental stress.

Field studies also can provide evidence to support theoretical results about the evolutionary consequences of inbreeding. The relationship between natural rates of self-fertilization and inbreeding depression can answer questions about their relative role in the evolution of breeding systems. Holtsford and Ellstrand examined how inbreeding effects vary with natural outcrossing rates in populations of Clarkia tembloriensis. The theory that inbreeding effects will be reduced in populations with high natural rates of inbreeding was not supported by their results; rather, all populations showed significant inbreeding depression for fecundity traits. In a similar study of Collinsia heterophylla, inbreeding depression effects also did not decline with as the level of natural inbreeding increased. However, these studies provide conflicting evidence for the role of purging of deleterious recessives from populations with high natural inbreeding rates.

Mechanisms to Avoid Inbreeding Depression

Plants display a number of variations in morphology and breeding systems that reduce the likelihood of inbreeding through selfing or mating with close relatives. However, avoidance of inbreeding is unlikely the only selective force leading to the evolution of these features. Many of the adaptations provide overt benefits to plant fitness beyond reducing selfing.

Varying the spatial distribution of male and female reproductive organs has evolved in many taxa of plants, and this breeding system can take on several different forms. Though most species have hermaphroditic flowers, some species have separate flowers with male and female functions. To reduce selfing, these flowers can be spatially distinct on the same individual plant (monoecy) or on separate individuals (dioecy). Many variations of these breeding systems exist, where plants may display combination of hermaphroditic and functionally male or female flowers. Plants also vary the spatial distribution of male and female organs within hermaphroditic flowers, which is known as heterostyly. This mechanism of self-pollination avoidance is present in 22 angiosperm families likely of independent evolutionary origin. In distylous species (e.g. primrose Primula vulgaris), two morphs of flowers are present: the 'pin' flowers with long styles and short stamens and the 'thrum' flowers with short styles and long stamens. The function of distyly in preventing inbreeding depends on interactions with the pollinators; pollen transfer between morphs rather than within morphs is facilitated by pollen deposition on different parts of the pollinator. Similarly, tristyly occurs when three flower morphs are present, e.g. the long-, mid- and short-styled morphs of purple loosestrife, Lythrum salicaria. Tristyly has most been studied in the context of its breakdown, where the advantage of self-fertilization outweighs the consequences of inbreeding depression.

Plants may also vary the distribution of male and female function in time, known collectively as dichogamy. Where pollen is dispersed from the anthers before the stigmas are receptive, the flowers are considered to be protandrous. Similarly, when the stigmas are receptive before pollen is shed from the anthers, the flowers are protogynous. Protandry is present in many more taxa than protogyny, indicating that dichogamy is mainly a means to control the amount of pollen exiting a flower or inflorescence.

Another means of inbreeding avoidance is via genetically mediated self-incompatibility, in which pollen with the same genotype as the maternal plant is unable to bring about fertilization. An example of this type of self-incompatibility is many apple cultivars that require other varieties to fertilize flowers for fruit to be produced. Rejection of pollen is controlled by a self-incompatibility locus known as the S-locus, where self-pollen and pollen from other plants with the same allele is rejected. Typically there are many different alleles at this locus in a population. There are two types of self-incompatibility in plants. Gametophytic incompatibility occurs where rejection is controlled by haploid genotype of the pollen itself. Sporophytic incompatibility occurs where rejection is controlled by the diploid genotype of the plant that produces the pollen.

Inbreeding and Conservation of Plants

Inbreeding is of concern to the conservation of rare plants because in many rare or threatened species high levels of inbreeding that occur in small, remnant populations. The level of inbreeding in a population is determined by a number of factors discussed above, but is ultimately determined by population size, as this will generally determine the diversity of available mates. Furthermore, human activities like selective logging can directly increase the level of inbreeding by increasing mating between relatives in wild populations of plants. The severity of inbreeding depression is overall less in populations that have regularly inbred in the past; for example, populations of narrowly endemic species with consistent small population sizes throughout their evolutionary history. However, many species that are currently endangered or threatened now occur in small populations, which may or may not reflect historical sizes. Inbreeding depression will proximately have detrimental consequences for population growth, but long-term population survival will ultimately depend on genetic variability. Thus, inbreeding depression is an issue in the viability of populations of rare or threatened plants.

Despite the significant negative effects of inbreeding depression demonstrated in studies from Darwin to today, the relative importance of inbreeding depression as a factor in population extinction is subject to debate. There is some doubt about the importance of genetic effects in determining the survival of small populations relative to demographic and environmental stochasticity. Recent studies have provided evidence for the role of inbreeding in population extinction by demonstrating the importance of genetic diversity. Genetic 'rescue' effects have been demonstrated in a metapopulation of Silene alba (white campion), where gene influx increased the persistence of isolated populations. Though direct empirical evidence of inbreeding depression contributing to population extinction is limited, other sources of information clearly suggest that the influence of genetic factors cannot be discounted. Demographic models have predicted that inbreeding depression can significantly effect population survival, especially when it impacts the survival of adults. Circumstantial evidence from recent studies also indicates that inbreeding depression plays a role in population survival. Heschel and Paige demonstrated increased offspring fitness when small, presumedly inbred, populations were outcrossed, but did not observe such increases when large populations were similarly treated. Menges found that seed germination increased with population size in isolated populations of a threatened prairie species, Silene regia (royal catchfly). Similarly, decreased fecundity and survival in relation to genetic diversity and population size were found in populations of Gentiana pneomonanthe (marsh gentian), a rare plant of European calcareous grasslands. However, a similar study of Senecio integrifolius (groundsel) failed to detect a clear relationship between inbreeding effects and population size, indicating that other factors were likely of primary importance for this plant.

POLYPLOIDY IN CROP IMPROVEMENT

An organism or individual having more than two basic or monoploid sets of chromosomes is called polyploid and such condition is known as polyploidy. It is estimated that about one third species of flowering plants are polyploids. In wild species of grass family polyploidy has been reported upto 70%.

Polyploidy is of two types, viz.:

- Autopolyploidy,
- Allopolyploidy.

Table: Type of polyploidy.

Type of polyploidy	Definition/brief description
Polyploidy	A condition with more than two genomes
• Autopolyploidy	Multiplication od the same genome
◦ Autotriploidy	Three copies of the same genome (3x)
◦ Autotetraploidy	Four copies of the same genome (4x)
◦ Autopentaploidy	Five copies of the same genome (5x)
◦ Autohexaploidy	Six copies of the same genome (6x)
• Allopolyploidy	A condition in which complete genomes of two or more species combine together.
◦ Allotetraploidy	Two copies each of two distinct genomes ($2x_1+2x_2$)
◦ Allohexaploidy	Two copies each of three distinct genomes ($2x_1+2x_2+2x_3$)
◦ Allooctaploidy	Two copies each of four distinct species ($2x_1+2x_2+2x_3+2x_4$)

Autopolyploidy

Polyploids which originate by multiplication of the chromosome of a single species are known as autopolyploids or autoploids and such situation is referred to as autopolyploidy. In other words, autoploidy refers to the situation in which additional sets of chromosomes arise from the same species.

Autoploids include triploids (3x), tetraploids (4x), pentaploids (5x), hexaploids (6x), septaploids (7x), octaploids (8x), and so on. Autoploids are also known as simple polyploids or single species polyploids.

Autotriploids

They have three sets of chromosomes of the same species. They can occur naturally or can be produced artificially by crossing between autotetraploid and diploid species. Triploids are generally highly sterile due to defective gamete formation. Triploids are useful only in those plant species which propagate asexually like banana, sugarcane, apple etc.

A few examples of practical uses of autotriploidy are given below:

Banana

Cultivated varieties of banana are triploids and seedless. Such bananas have larger fruits than diploid ones.

Apples

Some varieties of apple are triploids which are propagated asexually by budding or grafting.

Sugar-Beet

Triploid sugar-beets have higher sugar contents than diploids and are generally resistant to moulds.

Watermelon

Triploid watermelons are seedless or have rudimentary seeds like cucumber. These seedless watermelons are produced by crossing tetraploid female with diploid male. However, the reciprocal cross is not successful.

All other autoploids with odd chromosome sets, viz. pentaploids, septaploids, etc. also behave like triploids.

Autotetraploids

They have four copies of the genome of same species. They may arise spontaneously or can be induced artificially by doubling the chromosomes of a diploid species with colchicine treatment. Tetraploids are usually very stable and fertile because pairing partners are available during meiosis.

In such individuals diploid gametes (2n) are formed. Autotetraploids are usually larger and more vigorous than the diploid species, rye, grapes, alfalfa, groundnut, potato and coffee are well known examples of autotetraploids.

A few cases of practical use are described below:

Rye

Autotetraploid rye is grown in Sweden and Germany. They have larger seeds and higher proteins than diploids.

Grapes

Tetraploid grapes have been developed in California, USA, which have larger fruits and fewer seeds per fruit than diploids.

Alfalfa

Tetraploid varieties of alfalfa are better than diploid in yield and have better recovery after grazing.

Allopolyploidy

A ploypolid organism which originates by combining complete chromosome sets from two or more species is known as allopolyploid or alloploid and such condition is referred to as allopolyploidy. Alloploids are also known as hybrid polyploids or bispecies or multispecies polyploids.

An allopolyploid which arises by combining genomes of two diploid species is termed as allotetraploid or amphidiploid. Allopolyploidy can be developed by interspecific crosses and fertility is restored by chromosome doubling with colchicine treatment. Alloploidy has played greater role in crop evolution than autopolyploidy, because allopolyploidy is found in about 50% of crop plants.

Natural Alloploids

Some important natural allopolyploid crops are wheat, cotton, tobacco, mustard, oats etc. Interspecific crossing followed by chromosome doubling in nature have resulted in the origin of allopolyploids.

The origin of some natural allopolyploid crops is briefly presented below:

Wheat

The bread wheat (Triticum aestivum) is an allopolyploid. It is believed that A genome of wheat has come from Triticum monococcum (2n = 14), D genome from Triticum tauschi (2n = 14) and B genome from unknown source probably from an extinct species (2n = 14). Thus hexaploid wheat has two copies of the genomes from three species.

First allotetraploid Triticum turgidum developed from a cross between Triticum monococcum and unknown species of B genome. Then cross between T. turgidum and T. tauschi resulted in the development of hexaploid wheat T. aestivum.

Tobacco

There are two cultivated species of tobacco, viz. Nicotiana tabacum and N. rustica. N. tabacum is an amphidiploid between N. sylvestris (2n = 24) and N. tomentosa (2n = 24). N. rustica is believed to be amphidiploid between N. paniculata and N. undulata. Each of these two species has (2n = 24).

Cotton

The tetraploid American cotton (Gossypium hirsutum) is believed to be an amphidiploid between G. africanum, and G. raimondii. Both these species are diploid with 2n = 26. The chromosomes of G. africanum are larger than G. rainondii.

Oats

The cultivated oat (Avena sativa, n = 21) is an allohexaploid which is considered to have originated from a cross between A. barbata (tetraploid, n = 14) and A. strigosa (a diploid, n = 7).

Brassica

In brassica, there are there basic species, viz. Brassica nigra (BB, n = 8), B. oleracea (CC, n = 9) and B. campestris (AA, n = 10). The cross between B. nigra and B. oleracea gave rise to B. carinata. Cross between B. campestris and B. oleracea led to the development of B. napus, and cross between B. campestris and B. nigra resulted in the development of B. juncea. All the resulting species are amphidiploids.

Artificial Alloploids

Artificial alloploids have been synthesized in some crops with two main objectives, viz.:

- Either to study the origin of naturally available alloploids,
- To explore the possibilities of creating new species.

Some examples of artificial alloploids are given below:

Raphanobrassica

This is a classic example of artificially synthesized alloploid. This was developed between radish (Raphanus sativus, n = 9) and cabbage (Brassica oleracea, n = 9) by Russian geneticist Karpechenko in 1928. He wanted to develop a fertile hybrid between these two species with roots of radish and leaves of cabbage. But he got a fertile amphidiploid (4n = 36) by spontaneous chromosome doubling which unfortunately had roots of cabbage and leaves of radish. Thus it was of no use.

Tobacco

Clausen and Goodspeed synthesized a new hexaploid species of tobacco (Nicotiana) from a cross between Nicotiana tabacum (2n = 48) and N. glutinosa (2n = 24). The F_1 was sterile with 2n = 36, which was made fertile by doubling of chromosome through colchicine treatment. The new species is known as N. digluta.

Triticale

Triticale is a new crop species which has been synthesized from a cross between wheat (Triticum aesitvum) and rye (Secale cereale, n = 7). Some triticales are developed from cross between tetraploid wheat (Triticum turgidum) and rye and some from cross between hexaploid wheat (T. aestivum) and rye.

The F_1 was sterile, which was made fertile by colchicine treatment. Triticales produced using tetraploid and hexaploid wheat are hexaploid and octaploid, respectively. Titicale is now commonly grown in Canada, Mexico, Hungary and some other countries.

Wheat

In wheat, Mc Fadden and Sears developed hexaploid between Triticum turgidum (formerly T. dicoccum) and T. tauschi (formerly Aegilops squarrosa). This resembled T. aestivum (formerly T. spelta) and produced fertile F_1 when crossed with natural T. aestivum. This suggested involvement of these two species in the evolution of T. aestivum in the long past.

Cotton

The American upland cotton (Gossypium hirsutum) was synthesized from a cross between G. herbaceum and G. raimondii . Both these species are diploid with 2n – 26. The former is Old world cultivated diploid and the latter, New world wild diploid. This suggested involvement of these two species in the evolution of upland cotton.

Induction of Polyploidy

Polyploidy is mainly induced by treatment with a chemical known as colchicine. This is an alkaloid which is obtained from the seeds of a plant known as Colchicum autumnale, which belongs to the family Liliaceae. Colchicine does not affect Colchicum from which it is extracted, because this plant has an anti-colchicine substance.

Colchicine is applied in a very low concentration, because high concentration is highly toxic to the cells. For effective induction of polyploidy, usually concentrations of 0.01% to 0.5% are used in different plant species. The colchicine induced polyploidy is known as colchiploidy.

In plants colchicine is applied to growing tips, meristematic cells, seeds and axillary buds in aqueous solution or mixed with lanolin. The duration of treatment varies from 24 hours to 96 hours depending upon the species of plants.

Colchicine induces polyploidy by inhibiting formation of spindle fibres. The chromosomes do not line up on the equatorial plate and divide without moving to the poles due to lack of spindle fibres. The nuclear membrane is formed around them and the cell enters interphase. Thus nucleus has double the chromosome number.

Effects of Polyploidy

Polyploidy has marked effects on the morphology of plants. The distinct features of autoploids are increase in general vigour and size of various plant parts. Such features are generally referred to as gigantism.

Autoploids have the following important features:

- Stems are thicker and stouter.
- Leaves are fleshy, thicker, larger and deeper green in colour.
- Roots are stronger and longer.
- Flowers, pollens and seeds are larger than diploids.
- Maturity duration is longer and growth rate is slower than diploids.
- Water contents are higher than diploids etc.

Applications of Polyploidy in Crop Improvement

Polyploidy plays an important role in crop improvement. Both autopolyploidy and allopolyploidy are useful in several ways. However, allopolyploidy has wider applications than autopolyploidy.

Applications of autopolyploidy and allopolyploidy in crop improvement are briefly presented below:

Autoploidy

Both triploids and tetraploids have been used in crop improvement. However, their applications have been limited to few species only. Autotriploids have been developed in sugar-beets and watermelon. Triploid sugar-beets have larger roots and higher sugar contents than diploids.

Triploid watermelons are seedless or have rudimentary and soft seeds like cucumber. The triploid seed is produced by using tetraploid as female and diploid as male. The reciprocal cross is not successful. Moreover, triploid watermelons have irregular shape and fresh seeds have to be made every year.

Autotetraploids have been developed in forage crops like berseem, alfalfa and rye; vegetables like radish, turnip and cabbage; and fruits like grapes. Tetraploid varieties of rye are grown in Sweden and Germany. They have larger seeds and higher proteins than diploids.

Tetraploid grapes have been developed in California USA, which have larger fruits and fewer seeds per fruit than diploids. Tetraploid varieties of alfalfa are better than diploid in yield and recovery after grazing. However, tetraploid cabbage and turnips have higher water contents than diploids.

Alloploidy

Alloploidy is useful in four principal ways, viz.:

- In tracing the origin of natural allopolyploids,
- In creating new species,
- In interspecific gene transfer, and
- As a bridging species.

These are briefly described below:

Tracing the Origin of Crop Species

Alloploidy plays an important role in tracing the origin of natural allopolyploids. Study of chromosome pairing in a cross between allopolyploid and a diploid species helps in tracing the origin of polyploid species. The affinity in pairing indicates the involvement of diploid species in the evolution of such polyploid. Lack of pairing between the chromosomes of two species, rules out the involvement of diploid species, in the origin of polyploid under study.

Creation of New Species

Alloploidy sometimes leads to the creation of new crop species. Triticale is the best example, which is allopolyploid between wheat and rye. It combines desirable characters of both the species, i.e., grain quality of wheat and hardiness of rye. Triticales are of two types, viz. primary triticales and secondary triticales.

Primary triticales are derivatives of cross involving either tetraploid wheat or hexaploid wheat with rye. Secondary triticales are derivatives of the cross either between two primary triticales or between primary triticale and wheat. These are superior to primary triticales in several aspects.

Presently cultivated strawberry originated from a cross between North and South American species in the middle of seventeenth century in Europe. Loganberry was developed from a cross between raspberry and blackberry in 1880 in California, USA.

Interspecific Gene Transfer

When the desirable character is not found within the species, it is transferred from the related species. Interspecific gene transfer is done in two ways, viz. by alien addition and alien substitution. In case of alien addition one chromosome of wild species is added to the normal complement of a cultivated species.

In case of alien substitution, one pair of chromosome is substituted in cultivated species with those of wild donor species. Such type of gene transfer is generally made for disease resistance. This type of gene transfer has been achieved in crops like wheat, tobacco, cotton and oats. In cotton, lint strength has been transferred from G. thurberi to G. hirsutum.

However, in such gene transfer several undesirable characters (genes) are transferred along with desirable ones. In tobacco, mosaic resistance has been transferred from N. glutinosa to N. tabacum through alien addition.

Bridging Cross

Sometimes direct cross between two species is not possible due to sterility in F_1. In such case, first an amphidiploid is made between such species and then amphidiploid is crossed with the recipient species. Such bridging crosses have been made for transfer of genes from wild species particularly in crops like tobacco and cotton.

Besides these, alloploids have some other applications. For example, heterotic effects can be conserved more easily in allotetraploids than in diploids. Moreover, interspecific hybrids can be made fertile by artificial doubling of chromosomes.

Limitations of Polyploidy

Polyploidy has several limitations.

Some important limitations of polyploidy in crop improvement are briefly presented below:

Limited Use

The single species polyploidy has limited applications. It is generally useful in those crop species which propagate asexually like banana, potato, sugarcane, grapes etc.

Difficulty in Maintenance

The maintenance of monoploids and triploids is not possible in case of sexually propagating crop species.

Undesirable Characters

In bispecies or multispecies polyploids characters are contributed by each of the parental species. These characters may be sometimes undesirable as in case of Raphanobrassica.

Some Other Defects

Induced polyploids have several defects such as low fertility, genetic instability, slow growth rate, late maturity, etc.

Chances of developing new species through allopolyploidy are extremely low.

METHODS OF INSERTING FOREIGN DNA INTO CROP PLANTS

Plasmid Method

This is agrobacterium mediated technique of foreign gene transfer and is also known as biological method of gene transfer.

Main features of this method are presented below:

- It is used with cell culture or protoplast culture, hence regeneration from cells or protoplasts is a basic requirement.
- This is a cheaper method of transformation than gene gun method.
- It has some control over the copy number and site of integration of trans-genes.
- Main drawbacks of this method are host specificity, somaclonal variation and slow regeneration.

Particle Bombardment Method

In this method, foreign DNA is inserted into plant cells through gene gun using gold or tungsten coated DNA particles. It is also known as biolistic method, particle acceleration and micro ballistic method.

Main features of this method are presented below:

- It is used with meristems, immature embryos, embryo genic callus, suspension culture and hypocotyls.
- This method is simple than plasmid method and does not have host specificity.
- High cost of gene gun prohibits use of this method.
- There is no control over the copy number and site of foreign gene integration.

Direct DNA Transfer Method

This method involves electro proton and is used with protoplast. Hence regeneration from protoplast is basic requirement of this method. This method has limited applications.

Main features of this method are given below:

- It is used with protoplast culture.

- It has host specificity.
- The regeneration is slow.
- It is cheaper method of foreign gene transfer.

Micro-Injection Method

In this method, plasmid DNA is delivered into host cells by microscopic needles, known as micro-injection.

Main features of this method are presented below:

- It is used with protoplast culture.
- This has no host specificity.
- The regeneration is slow from protoplast.
- It is inexpensive method of foreign gene transfer.

Out of these, first two methods, viz., plasmid method and gene gun method are widely used for development of transgenic crop plants. Other two methods are not in much use.

GENE DELIVERY METHODS IN PLANTS

Cells or protoplasts can be cultured and used for regeneration of whole plants. Therefore, these protoplasts can be used for gene transfer followed by regeneration leading to the production of transgenic plants. Besides cultured cells and protoplasts, other meristem cells (immature embryos or organs), pollen or zygotes can also be used for gene transfer in plants. The enormous diversity of plant species and the availability of diverse genotypes in a species, made it necessary to develop a variety of techniques, suiting different situations. These different methods of gene transfer in plants are discussed.

Target Cells for Gene Transformation

The first step in gene transfer technology is to select cells that are capable of giving rise to whole transformed plants. Transformation without regeneration and regeneration without transformation are of limited value. In many species, identification of these cell types is difficult. This is unlike the situation in animals, because the plant cells are totipotent and can be stimulated to regenerate into whole plants in vitro via organogenesis or embryogenesis. However, in vitro plant regeneration imposes a degree of 'genome stress', especially if plants are regenerated via a callus phase. This may lead to chromosomal or genetic abnormalities in regenerated plants a phenomenon referred to as soma clonal variation.

In contrast to this, gene transfer into pollen (or possibly egg cells) may give rise to genetically transformed gametes, which if used for fertilization (in vivo) may give rise to transformed whole plants. Similarly, insertion of DNA into zygote (in vivo or in vitro) followed by embryo rescue, may also be used to produce transgenic plants. Another alternative approach is the use of individual cells in embryos or meristems, which may be grown in vitro or may be allowed to develop normally for the production of transgenic plants.

Vectors for Gene Transfer

Most vectors carry marker genes, which allow recognition of transformed cells (other cells die due to the action of an antibiotic or herbicide) and are described as selectable markers. Among these marker genes, the most common selectable marker is npt II, providing kanamycin resistance. Other common features of suitable transformation vector include the following: (i) multiple unique restriction sites (a synthetic polylinker); (ii) bacterial origins of replication (e.g. ColE1).

The vectors having these properties may not necessarily have features, which facilitate their transfer to plant cells or integration into the plant nuclear genome. Therefore, Agrobacterium Ti plasmid is preferred over all other vectors, because of wide host range of this bacterial system and the capacity to transfer genes due to the presence of T - DNA border sequences.

Gene Delivery Methods

To achieve genetic transformation in plants, we need the construction of a vector (genetic vehicle) which transports the genes of interest, flanked by the necessary controlling sequences i.e. promoter and terminator, and deliver the genes into the host plant. The two kinds of gene transfer methods in plants.

Vector-Mediated or Indirect Gene Transfer

Among the various vectors used in plant transformation, the Ti plasmid of Agrobacterium tumefaciens has been widely used. This bacterium is known as "natural genetic engineer" of plants because these bacteria have natural ability to transfer T-DNA of their plasmids into plant genome upon infection of cells at the wound site and cause an unorganized growth of a cell mass known as crown gall. Ti plasmids are used as gene vectors for delivering useful foreign genes into target plant cells and tissues. The foreign gene is cloned in the T-DNA region of Ti-plasmid in place of unwanted sequences. To transform plants, leaf discs (in case of dicots) or embryogenic callus (in case of monocots) are collected and infected with Agrobacterium carrying recombinant disarmed Ti-plasmid vector. The infected tissue is then cultured (co-cultivation) on shoot regeneration medium for 2-3 days during which time the transfer of T-DNA along with foreign genes takes place. After this, the transformed tissues (leaf discs/calli) are transferred onto selection cum plant regeneration medium supplemented with usually lethal concentration of an antibiotic to selectively eliminate non-transformed tissues. After 3-5 weeks, the regenerated shoots (from leaf discs) are transferred to root-inducing medium, and after another 3-4 weeks, complete plants are transferred to soil following the hardening (acclimatization) of regenerated plants. The molecular techniques like PCR and southern hybridization are used to detect the presence of foreign genes in the transgenic plants.

Structure and Functions of Ti and Ri Plasmids

The most commonly used vectors for gene transfer in higher plants are based on tumour inducing mechanism of the soil bacterium Agrobacterium tumefaciens, which is the causal organism for crown gall disease, A closely related species A. rhizogenes causes hairy root disease. An understanding of the molecular basis of these diseases led to the utilization of these bacteria for developing gene transfer systems. It has been shown that the disease is caused due to the transfer of a DNA segment from the bacterium to the plant nuclear genome. The DNA segment, which is transferred is called T - DNA and is part of a large Ti (tumour inducing) plasmid found in virulent strains of Agrobacterium tumefaciens. Similarly Ri (root inducing) megaplasmids are found in the virulent strains of A. rhizogenes.

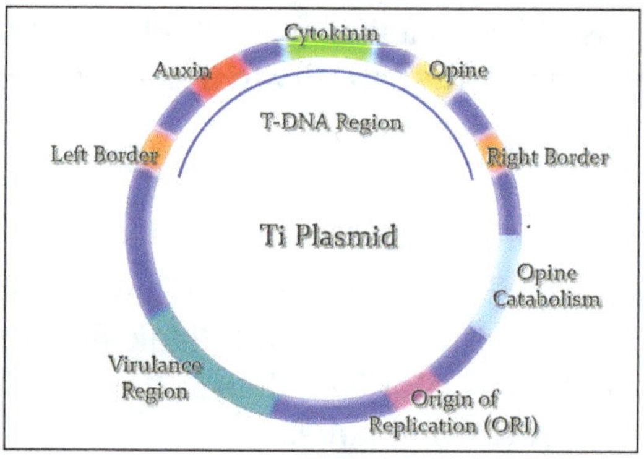

Most Ti plasmids have four regions in common, (i) Region A, comprising T-DNA is responsible for tumour induction, so that mutations in this region lead to the production of tumours with altered morphology (shooty or rooty mutant galls). Sequences homologous to this region are always transferred to plant nuclear genome, so that the region is described as T-DNA (transferred DNA). (ii) Region B is responsible for replication. (iii) Region C is responsible for conjugation. (iv) Region D is responsible for virulence, so that mutation in this region abolishes virulence. This region is therefore called virulence (v) region and plays a crucial role in the transfer of T-DNA into the plant nuclear genome. The components of this Ti plasmid have been used for developing efficient plant transformation vectors.

The T-DNA consists of the following regions: (i) An one region consisting of three genes (two genes tms and tms2 representing 'shooty locus' and one gene tmr representing 'rooty locus') responsible for the biosynthesis of two phytohormones, namely indole acetic acid (an auxin) and isopentyladenosine 5'-monophosphate (a cytokinin). These genes encode the enzymes responsible for the synthesis of these phytohormones, so that the incorporation of these genes in plant nuclear genome leads to the synthesis of these phytohormones in the host plant. The phytohormones in their turn alter the developmental programme, leading to the formation of crown gall (ii) An os region responsible for the synthesis of unusual amino acid or sugar derivatives, which are collectively given the name opines. Opines are derived from a variety of compounds (e.g. arginine + pyruvate), that are found in plant cells. Two most common opines are octopine and nopaline. For the synthesis of octopine and nopaline, the corresponding enzymes octopine synthase and nopaline synthase are coded by T- DNA.

Depending upon whether the Ti plasmid encodes octopine or nopaline, it is described as octopine-type Ti plasmid or nopalinetype Ti plasmid. Many organisms including higher plants are incapable of utilizing opines, which can be effectively utilized by Agrobacterium. Outside the T-DNA region, Ti plasmid carries genes that, catabolize the opines, which are utilized as a source of carbon and nitrogen. The T-DNA regions on all Ti and Ri plasmids are flanked by almost perfect 25bp direct repeat sequences, which are essential for T-DNA transfer, acting only in cis orientation. It has also been shown that any DNA sequence, flanked by these 25bp repeat sequences in the correct orientation, can be transferred to plant cells, an attribute that has been successfully utilized for Agrobacterium mediated gene transfer in higher plants leading to the production of transgenic plants.

Besides 25bp flanking border sequences (with T DNA), vir region is also essential for T-DNA transfer. While border sequences function in cis orientation with respect to T - DNA, vir region is capable of functioning even in trans orientation. Consequently physical separation of T-DNA and vir region onto two different plasmids does not affect T-DNA transfer, provided both the plasmids are present in the same Agrobacterium cell. This property played an important role in designing the vectors for gene transfer in higher plants, as will be discussed later. The vir region (approx 35 kbp) is organized into six operons, namely vir A, vir B, vir C, vir D, vir E, and vir G, of which four operons (except vir A and vir G) are polycistronic. Genes vir A, B, D, and G are absolutely required for virulence; the remaining two genes vir C and E are required for tumour formation. The vir A locus is expressed constitutively under all conditions.

The vir G locus is expressed at low levels in vegetative cells, but is rapidly induced to higher expression levels by exudates from wounded plant tissue. The vir A and vir G gene products regulate the expression of other vir loci. The vir A product (Vir A) is located on the inner membrane of Agrobacterium cells and is probably a chemoreceptor, which senses the presence of phenolic compounds (found in exudates of wounded plant tissue), such as acetosyringone and β-hydroxyaceto syringone. Signal transduction proceeds via activation (possibly phosphorylation) of Vir G (product of gene vir G), which in its turn induces expression of other vir genes.

Transformation Techniques using Agrobacterium

Agrobacterium infection (utilizing its plasmids as vectors) has been extensively utilized for transfer of foreign DNA into a number of dicotyledonous species. The only important species that have not responded well, are major seed legumes, even though transgenic soybean (Glycine max) plants have been obtained. The success in this approach for gene transfer has resulted from improvement in tissue culture technology. However, monocotyledons could not be successfully utilized for Agrobacterium mediated gene transfer except a solitary example of Asparagus. The reasons for this are not fully understood, because T-DNA transfer does occur at the cellular level. It is possible that the failure in monocots lies in the lack of wound response of monocotyledonous cells.

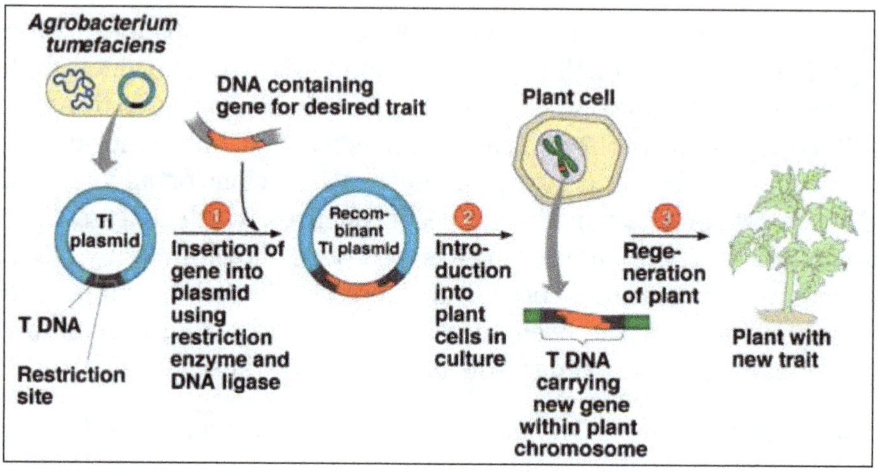

Vectorless or Direct Gene Transfer

In the direct gene transfer methods, the foreign gene of interest is delivered into the host plant cell without the help of a vector. The gene transfer system using genetically engineered vectors do not work out well particularly in monocot species. Considering the problem, direct gene transfer methods have been tried and the methods used for direct gene transfer in plants.

Chemical Mediated Gene Transfer

Direct DNA uptake by protoplasts can be stimulated by chemicals like polyethylene glycol (PEG). This method was reported by Krens and his colleagues in 1982. The technique is so efficient that virtually every protoplast system has proven transformable. PEG is also used to stimulate the uptake of liposomes and to improve the efficiency of electroporation. PEG at high concentration (15-25%) will precipitate ionic macromolecules such as DNA and stimulate their uptake by endocytosis without any gross damage to protoplasts. This is followed by cell wall formation and initiation of cell division. These cells can now be plated at low density on selection medium. Initial studies using the above method were restricted to Petunia and Nicotiana.

However, other plant systems (rice, maize, etc.) were also successfully used later. In these methods, PEG was used in combination with pure Ti plasmid, or calcium phosphate precipitated Ti plasmid mixed with a carrier DNA. Transformation frequencies upto 1 in 100 have been achieved by this method. Nevertheless, there are serious problems in using this method for getting transgenic plants and all these problems relate to plant regeneration from protoplasts.

Microinjection and Macroinjection

Plant regeneration from transformed protoplasts, still remains a problem. Therefore cultured tissues, that encourage the continued development of immature structures, provide alternate cellular targets for transformation. These immature structures may include immature embryos, meristems, immature pollen, germinating pollen, isolated ovules, embryogenic suspension cultured cells, etc. The main disadvantage of this technique is the production of chimeric plants with only a part of the plant transformed. However, from this chimeric plant, transformed plants of single cell origin can be subsequently obtained. Utilizing this approach, transgenic chimeras have actually been obtained in oilseed rape (Brassica napus).

When cells or protoplasts are used as targets in the technique of microinjection, glass micropipettes with 0.5-10μm diameter tip are used for transfer of macromolecules into the cytoplasm or the nucleus of a recipient cell or protoplast. The recipient cells are immobilized on a solid support (cover slip or slide, etc.) or artificially bound to a substrate or held by a pipette under suction (as done in animal systems). Often a specially designed micromanipulator is employed for microinjecting the DNA. Although, this technique gives high rate of success, the process is slow, expensive and requires highly skilled and experienced personnel.

The microinjection method was introduced by two groups of scientist led by Crossway and Reich in l986. Recently a method known as “holding pipette method” was introduced. In this, the protoplasts are isolated from cell suspension culture are placed on a depression slide, by its side with a microdroplet of DNA solution. Using the holding pipette, the protoplast has to be held and the DNA to be injected into the nucleus using the injection pipette. After the micro injection the injected cells are cultured by hanging droplet culture method.

DNA macroinjection employing needles with diameters greater than cell diameter has also been tried. In rye (Secale cereale), a marker gene was macroinjected into the stem below the immature floral meristem, so as to reach the sporogenous tissue leading to successful production of transgenic plants. Unfortunately, this technique could not be successfully repeated with any other cereal, when tried in several laboratories. Therefore, doubt is expressed about the validity of earlier experiments conducted with rye.

Electroporation Method

Electroporation is another efficient method for the incorporation of foreign DNA into protoplasts, and thus for direct gene transfer into plants. This method was introduced by Fromm and his coworkers in 1986.

This method is based on the use of short electrical impulses of high field strength. These impulses increase the permeability of protoplast membrane and facilitate entry of DNA molecules into the cells, if the DNA is in direct contact with the membrane. In view of this, for delivery of DNA to protoplasts, electroporation is one of the several routine techniques for efficient transformation. However, since regeneration from protoplasts is not always possible, cultured cells or tissue explants are often used. Consequently, it is important to test whether electroporation could transfer genes into walled cells. In most of these cases no proof of transformation was available.

The electroporation pulse is generated by discharging a capacitor across the electrodes in a specially designed electroporation chamber. Either a high voltage (1.5 kV) rectangular wave pulse of short duration or a low voltage (350V) pulse of long duration is used. The latter can be generated by a home made machine. Protoplasts in an ionic solution containing the vector DNA are suspended between the electrodes, electroporated and then plated as usual. Transformed colonies are selected as described earlier. Using electroporation method, successful transfer of genes was achieved with the protoplasts of tobacco, petunia, maize, rice, wheat and sorghum. In most of these cases cat gene associated with a suitable promoter sequence was transferred. Transformation frequencies can be further improved by (i) using field strength of 1.25kV/cm, (ii) adding PEG after adding DNA, (iii) heat shocking protoplasts at 45 °C for 5 minutes before adding DNA and (iv) by using linear instead of circular DNA.

Microprojectiles or Biolistics or Particle Gun for Gene Transfer

In 1987, Klein and his colleagues evolved a method by which the delivery of DNA into cells of intact plant organs or cultured cells is done by a process called Projectile Bombardment. The micro-projectiles (small high density particles) are accelerated to high velocity by a particle gun apparatus. These particles with high kinetic energy penetrate the cells and membranes and carry foreign DNA inside of the bombarded cells. This method is otherwise called as "Biolistics Method". In recent years, it has been shown that DNA delivery to plant cells is also possible, when heavy microparticles (tungsten or gold) coated with the DNA of interest are accelerated to a very high initial velocity (1,400 ft per, sec). These microprojectiles, normally 1-3pm in diameter, are carried by a 'macroprojectile' or the 'bullet' and are accelerated into living plant cells (target cells can be pollen, cultured cells, cells in differentiated tissues and meristems) so that they can penetrate cell walls of intact tissue. The acceleration is achieved either by an explosive charge (cordite explosion) or by using shock waves initiated by a high voltage electric discharge. The design of two particle guns used for acceleration of microprojectiles.

Gene gun.

Transformed plants using the above technique have been obtained in many cases including soybean, tobacco, maize, rice, wheat, etc. Transient expression of genes transferred in cells by this method has also been observed in onion, maize, rice and wheat. There is no other gene transfer approach, which has met with so much of enthusiasm. Consequently considerable investment has been made in experimentation and manpower for development of this technique.

Sonication Method: This is a simple technique recently formulated by Xu and his coworkers. In this method the explants (especially leaves) are excised and cut into segments, immmersed in sonication buffer containing plasmid DNA and Carrier DNA in a sterile glass petridish. Then the samples were sonicated with an ultrasonic pulse generator at 0.5 c/cm2 acoustic intensity for 30 minutes. After 30 minutes, the explants were rinsed in buffer solution without DMSO and transferred to the culture medium.

Transformation

This method is used for introducing foreign DNA into bacterial cells e.g. E. Coli. The transformation frequency (the fraction of cell population that can be transferred) is very good in this method. E.g. the uptake of plasmid DNA by E. coli is carried out in ice cold $CaCl_2$ (0-5 °C) followed by heat shock treatment at 37-45 °C for about 90 sec. The transformation efficiency refers to the number of transformants per microgram of added DNA. The $CaCl_2$ breaks the cell wall at certain regions and binds the DNA to the cell surface.

Conjuction

It is a natural microbial recombination process and is used as a method for gene transfer. In conjuction, two live bacteria come together and the single stranded DNA is transferred via cytoplasmic bridges from the donor bacteria to the recipient bacteria.

Liposome Mediated Gene Transfer or Lipofection

Liposomes are small lipid bags, in which large number of plasmids are enclosed. They can be induced to fuse with protoplasts using devices like PEG, and therefore have been used for gene transfer. The technique, offers following advantages: (i) protection of DNA/RNA from nuclease digestion, (ii) low cell toxicity, (iii) stability and storage of nucleic acids due to encapsulation in liposomes, (iv) high degree of reproducibility and (v) applicability to a wide range of cell types. In this technique, DNA enters the protoplasts due to endocytosis of liposomes, involving the following steps: (i) adhesion of the liposomes to the protoplast surface, (ii) fusion of liposomes at the site of adhesion and (iii) release of plasmids inside the cell. The technique has been successfully used to deliver DNA into the protoplasts of a number of plant species (e.g. tobacco, petunia, carrot, etc.).

Gene Transformation using Pollen or Pollen Tube

There has been a hope that DNA can be taken up by the germinating pollen and can either integrate into sperm nuclei or reach the zygote through the pollen tube pathway. Both these approaches have been tried and interesting phenotypic alterations suggesting gene transfer have been obtained. In no case, however, unequivocal proof of gene transfer has been available. In a number of experiments, when marker genes were used for transfer, only negative results were obtained. Several problems exist in this method and these include the presence of cell wall, nucleases, heterochromatic state of acceptor DNA, callose plugs in pollen tube, etc. Transgenic plants have never been recovered using this approach and this method, though very attractive, seems to have little potential for gene transfer.

Calcium Phosphate Precipitation Method for Gene Transfer

Foreign DNA can also be carried with the Ca^{++} ions, to be released inside the cell due to the precipitation of calcium in the form of calcium phosphate. In the past, this method was considered to be very important for gene transfer in plants.

Incubation of Dry Seeds, Embryos, Tissues or Cells in DNA

Incubation of dry seeds, embryos, tissues or cells in known DNA (viral or non viral having defined marker genes) has been tried in many cases and expression of defined genes has been witnessed. However, in no case proof of integrative transformation could be available. In all these cases, plant cell walls not only work as efficient barriers, but are also efficient traps for DNA molecules. It would be very surprising if DNA can cross cell walls efficiently without permeabilizing them either by PEG, or by electroporation or by any other device.

Selection of Transformed Cells from Untransformed Cells

The selection of transformed plant cells from untransformed cells is an important step in the plant genetic engineering. For this, a marker gene (e.g. for antibiotic resistance) is introduced into the plant along with the transgene followed by the selection of an appropriate selection medium (containing the antibiotic). The segregation and stability of the transgene integration and expression in the subsequent generations can be studied by genetic and molecular analyses (Northern, Southern, Western blot, PCR).

Though several methods have been described for gene transfer using naked DNA, the recovery of genetic recombinants, otherwise called as "transgenic plants" appears to be a rare phenomenon. A concerted effort:

- To accurately identify genes which can be shown to influence agronomically important characters,
- To apply the technique to clone the isolated genes,
- To anneal them to appropriate vectors,
- To evaluate their expression in agronomically important crop varieties will solve the deficiencies in the conventional breeding procedures.

The last five years have seen successful outcomes and transgenic plants have been produced in some crop species by using both vector mediated and direct gene transfer techniques. However, all the programmes were not successful because of lack of proofs for the integrative gene transfer. Considering the needs to have integrative gene transfer, Potrykus points out that all the successful gene transfers should have the following proofs:

- Serious controls for treatments and analysis.
- A tight correlation between treatment and predicted results.
- A tight correlation between physical (Southern blot, in situ hybridization) and phenotypic data.
- Complete Southern Analysis to show the hybrid fragments of host DNA and foreign DNA, and the absence or presence of contaminating fragments.
- Data that allow discrimination between false positives and correct transformants in the evaluation of the phenotypic evidence.
- Correlation of the physical and phenotypic evidence with transmission to sexual offspring, as well as genetic and molecular analysis of offspring populations.

References

- Crop-genetics, about-crop-science: crops.org, Retrieved 30 August, 2019
- Schweitzer, J.A.; Martinsen, G.D.; Whitham, T.G. (2002). "Cottonwood hybrids gain fitness traits of both parents: a mechanism for their long-term persistence?". American Journal of Botany. 89 (6): 981–990. Doi:10.3732/ajb.89.6.981

- Principles-of-Plant-Biotechnology, uploads, wp-content: agrimoon.com, Retrieved 29 March, 2019
- Frisch M, Melchinger AE (2005). "Selection theory for marker-assisted backcrossing". Genetics. 170 (2): 909–17. Doi:10.1534/genetics.104.035451. PMC 1450430. PMID 15802512
- Default, pocket, publications, resources: isaaa.org, Retrieved 2 June, 2019
- (2014) Plant Breeding and Genetics Joint FAO/IAEA Division of Nuclear Techniques in Food and Agriculture, Retrieved 31 July 2014
- Inserting-foreign-dna-into-crop-plant-genetics, transgenic-technology, biotechnology: biology-discussion.com, Retrieved 23 February, 2019
- Kaskey, Jack (21 November 2013) The Scariest Veggies of Them allbloomberg Business Week, Retrieved 31 July 2014
- Polyploidy-types-and-effects-methods-crop-improvement-botany, polyploidy, breeding-methods, plant-breeding-2: botanylibrary.com, Retrieved 22 May, 2019

Crop Production Systems

CHAPTER 3

On the basis of the type of crop and use, the agricultural production systems are classified into various practices such as mixed farming, subsistence farming, plantation farming, shifting cultivation, intercropping, monocropping, companion planting, etc. This chapter discusses in detail these crop production systems.

CROP PRODUCTION

Crop production is a branch of agriculture that deals with growing crops for use as food and fiber. Crop production includes grains, cotton, tobacco, fruits, vegetables, nuts and plants. Different crops grow best in different areas of the country. Warmer climates are ideal for growing citrus crops; northern states are best for growing apples and blueberries and the Midwest is ideal for growing grains, including wheat.

There are different types of agricultural production system depending on the type of crop and use. The types of feed or row crop grown by farmers depends on the traditional, organic or conventional management systems available. Crop Production and management of corn, cotton,wheat, soybean and tobacco crops generates profit to the farmers. Crop production also includes feed sources and resource inputs used to produce crops required to maintain the dairy herd and contribute to the meat industry. Feeds grown include corn grain or silage; alfalfa hay and silage; soybean and soybean meal; oats; wheat; distiller's grains solids. Animals maintained by farmers are provided with dietary supplements or minerals and grass or hay for forage. Resource inputs in crop production include equipments or machinery, fuel for tractors, pesticides, fertilizers and packaging materials.

Types of System

There are several hybrid crop production systems arising due to availability of certain natural resources and other factors.

Some of food crop production practices include mixed, subsistence, plantation farming and others. Mixed farming is an agricultural system practiced on the same piece of land by farmers to cultivate crops and raise animals simultaneously. Different crops with different maturity periods are grown continuously throughout the season at same time using best practices with good rainfall or irrigation facilities. Subsistence farming allows farmer to produce food with simple farm tools on small land holding. The farmers in this system are perceived poor that do not use electricity and irrigation system, fertilisers, pesticides or improved seeds;reducing the productivity.

Plantation farming or tree crop farming is an agricultural farming system for farmers of single crop like cocoa, tea, coffee, rubber, spices or fruits like apples, avocado, grapes, oranges, mangoes, etc grown on commercial basis on a large piece of land. The system requires good management and technical skills with a substantial amount of capital investment for machines, fertilizers and other facilities. In shifting cultivation system, farmers clear a piece of forest land by cutting and burning the residues of vegetation to grow crops for three to five years. The farmer abandons the land for a fallow period on loss of fertility and moves to a new fertile land to cultivate. The process is repeated and former lands are cultivated after years on regaining fertility. Arable system of farming is practiced in a small scale or commercial scale to grow only annual crops like cassava, plantain, vegetables, grains and legumes without mixed or pastoral farming.

The most common types of feed production involve nomadic and pastoral farming. Pastoral or Livestock farming avoid crops and aim at producing only livestock for dairy farming, raising beef cattle or sheep for wool. Farmers use available feed resources on established pasture lands to feed the livestock without moving around like in nomadic farming. This system is expensive and not sustainable when excessive grazing destroys the natural fields forcing farmers to buy feed for the herd. Agricultural farming system of nomadic farming is similar to pastoral farming but herdsmen move animals like cattle, goats, sheep, horses, camel or donkeys in search of water and suitable grazing fields.

MIXED FARMING

Mixed farming is an agricultural system in which a farmer conducts different agricultural practice together, such as cash crops and livestock. The aim is to increase income through different sources and to complement land and labour demands across the year.

Mixed farming systems can be classified in many ways. They can be based on land size, type of crops and animals, geographical distribution, market orientation, and so on. Three major categories are distinguished here:

- On-farm versus Between-farm Mixing: On-farm mixing refers to mixing on the same farm, and between-farm mixing refers to exchanging resources between different farms. On-farm mixing enables the recycling of resources generated on a single farm. Between-farm mixing can be used to resolve waste disposal problems where by crop farmers use dung from animal farms for fertiliser.

- Mixing within Crops and/or Animal Systems: This practice involves multiple cropping or keeping different types of animals together. For example, grain-legume association can provide grain with nitrogen. With plant inter-cropping farmers can make the most of the space available to them by selecting plants and cropping formations that maximise the advantage of light, moisture and soil nutrients. Examples of mixed animal systems include chicken-fish production where chicken dung serves as fish fodder.
- Diversified versus Integrated Systems: In a diversified system some components exist as independent units. In an integrated system, maximum use is made of resources, making the system highly interdependent.

Mixed farming technology contributes to adaptation to climatic change because the diversification of crops and livestock allows farmers to have a greater number of options to face the uncertain weather conditions associated with the increased climate variability. Mixed farming can also give a more stable production because if one crop or variety fails, another may compensate. Livestock represents a means by which families can save and invest in the future. Livestock is a walking bank of assets that can be sold during periods of need such as if crops fail due to drought or flooding.

Advantages of the Technology

This technology also allows greater food security and improved household nutrition levels. In addition, farmers can generate a surplus of some products that can be sold at market. Among other benefits, this technology also allows farmers to grow fodder for livestock and poultry. An additional benefit of mixed rice–fish culture systems is that the fish may help reduce populations of existing and emerging disease vectors such as mosquitoes.

In many areas the hungry season on farms comes in the months just after the rains start when producers need to invest labour in the planting and management of crops, but before they start to produce. Conversely grass production starts immediately with the rains, and livestock quickly gain weight and increase milk production. The high milk production in the rains can greatly help support the nutrition of farmers while they are tending their crops and waiting for harvest.

Advantages of Mixed Farming Systems for the Environment

- Mixed farming systems maintain soil fertility by recycling soil nutrients and allowing the introduction and use of rotations between various crops and forage legumes and trees, or for land to remain fallow and grasses and shrubs to become re-established;
- Mixed farming systems maintain soil biodiversity, minimise soil erosion, help to conserve water and provide suitable habitats for birds;

- Mixed farming systems make the best use of crop residues. When they are not used as feed, stalks may be incorporated directly into the soil, where, for some time, they act as a nitrogen trap, exacerbating deficiencies. Burning the crop residues, the other alternative, increases carbon dioxide emissions;
- Mixed farming systems allow intensified farming, with less dependence on natural resources and preserving more biodiversity than would be the case if food demands were to be met by crop and livestock activities undertaken in isolation.

Disadvantages of the Technology

One limitation is that production levels in mixed systems (tons per hectare, milk per animal daily, increase and reproduction rates), can be lower than in specialised systems (monoculture). Another disadvantage is that where farmers depend on wild animal stock rather than domesticated species, they may face increased vulnerability in instances where animal population levels are affected due to climate change (for example, where livestock populations need to be trimmed).

Partly because of overgrazing, some mixed farming systems of the tropical highlands of Asia and Central Africa are among the most eroded and degraded systems of the world. Integrating crops and livestock can help improve soil neutirent and reduce the stress on farming land.

Financial Requirements and Costs

As for most cases, to estimate the costs of implementing this technology the cost of wages, agricultural tools, and inputs (such as seeds and fertilisers) must be considered. Infrastructure for supporting livestock will be an added cost in crop-animal systems. The main financial needs are associated with credits for the acquisition of inputs, investment in training and in the dissemination of this technology. Investment is needed also to obtain the necessary qualitative and quantitative micro-climate information for managing the synchronisation of mixed crop cycles (phenologies).

Institutional and Organisational Requirements

Knowledge about the phenology and lifecycles of different crops to be cultivated simultaneously (or with some time lag between them), as well as knowledge about the soil nutrients required by each of them (so they can complement each other, is required to use this technology. For example, some crops require higher levels of nitrogen, while others require more potassium. Knowledge is also required about the season in which each of the crops involved should be sown and which crop species and varieties can be grown together because of the inter-species competition.

The organisations promoting this technology need to have qualified technicians both in agronomy and livestock production. These organisations must identify the farmers that are familiar with the technique of multiple crops in the area and develop positive relationships with them.

Barriers to Implementation

The main obstacle for the implementation of this technology is farmers' reluctance since mixed farming is considered to have low productivity in comparison with monocultures which have a high yield in terms of tons per hectare (t/ha). The best way to overcome these barriers is to demonstrate mixed farming systems with better productivity levels; to disseminate the benefits of this technology, and to provide training.

Opportunities for Implementation

The main opportunity for implementing mixed farming is that it improves and guarantees the range of products a farmer has available to sell at market. One option to increase productivity while maintaining economic and environmental benefits of mixed farming is specialisation. Partnerships with specialised farms are formed to facilitate the exchange of crops and waste products from manure. For example, the traditional association between nomads and farmers reaping where nomadic cattle converts crop residues into manure for cultivation. More recent developments include partnerships between dairy farmers and vegetable growers. Similarly, in organic farming in Europe between specialised organic farms there is an exchange of secondary products and crop residues for manure.

SHIFTING CULTIVATION

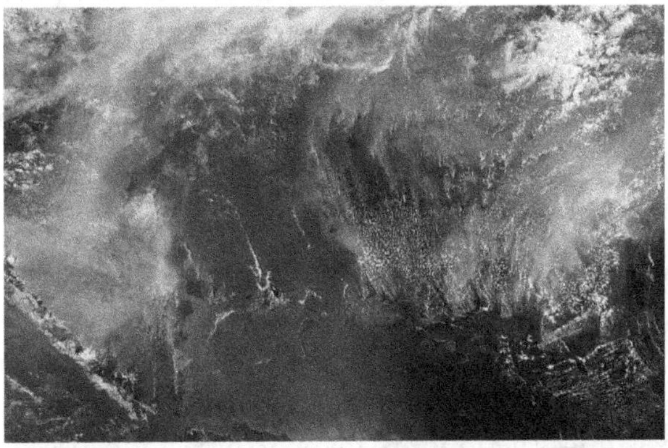

Slash-and-burn based shifting cultivation is a widespread historical practice in southeast Asia. Above is a satellite image of Sumatra and Borneo showing shift cultivation fires from October 2006.

Shifting cultivation is an agricultural system in which plots of land are cultivated temporarily, then abandoned and allowed to revert to their natural vegetation while the cultivator moves on to another plot. The period of cultivation is usually terminated when the soil shows signs of exhaustion or, more commonly, when the field is overrun by weeds. The length of time that a field is cultivated is usually shorter than the period over which the land is allowed to regenerate by lying fallow. This technique is often used in LEDCs (Less Economically Developed Countries) or LICs (Low Income Countries). In some areas, cultivators use a practice of slash-and-burn as one element of their farming cycle. Others employ land clearing without any burning, and some cultivators are purely migratory and do not use any cyclical method on a given plot. Sometimes no slashing at all is needed where regrowth is purely of grasses, an outcome not uncommon when soils are near exhaustion and need to lie fallow. In shifting agriculture, after two or three years of producing vegetable and grain crops on cleared land, the migrants abandon it for another plot. Land is often cleared by slash-and-burn methods—trees, bushes and forests are cleared by slashing, and the remaining vegetation is burnt. The ashes add potash to the soil. Then the seeds are sown after the rains.

Political Ecology

Shifting cultivation is a form of agriculture or a cultivation system, in which, at any particular point in time, a minority of 'fields' are in cultivation and a majority are in various stages of natural re-growth. Over time, fields are cultivated for a relatively short time, and allowed to recover, or are fallowed, for a relatively long time. Eventually a previously cultivated field will be cleared of the natural vegetation and planted in crops again. Fields in established and stable shifting cultivation systems are cultivated and fallowed cyclically. This type of farming is called jhumming in India.

Fallow fields are not unproductive. During the fallow period, shifting cultivators use the successive vegetation species widely for timber for fencing and construction, firewood, thatching, ropes, clothing, tools, carrying devices and medicines. It is common for fruit and nut trees to be planted in fallow fields to the extent that parts of some fallows are in fact orchards. Soil-enhancing shrub or tree species may be planted or protected from slashing or burning in fallows. Many of these species have been shown to fix nitrogen. Fallows commonly contain plants that attract birds and animals and are important for hunting. But perhaps most importantly, tree fallows protect soil against physical erosion and draw nutrients to the surface from deep in the soil profile.

The relationship between the time the land is cultivated and the time it is fallowed are critical to the stability of shifting cultivation systems. These parameters determine whether or not the shifting cultivation system as a whole suffers a net loss of nutrients over time. A system in which there is a net loss of nutrients with each cycle will eventually lead to a degradation of resources unless actions are taken to arrest the losses. In some cases soil can be irreversibly exhausted (including erosion as well as nutrient loss) in less than a decade.

The longer a field is cropped, the greater the loss of soil organic matter, cation-exchange-capacity and in nitrogen and phosphorus, the greater the increase in acidity, the more likely soil porosity and infiltration capacity is reduced and the greater the loss of seeds of naturally occurring plant species from soil seed banks. In a stable shifting cultivation system, the fallow is long enough for the natural vegetation to recover to the state that it was in before it was cleared, and for the soil to recover to the condition it was in before cropping began. During fallow periods soil temperatures are lower, wind and water erosion is much reduced, nutrient cycling becomes closed again, nutrients are extracted from the subsoil, soil fauna decreases, acidity is reduced, soil structure, texture and moisture characteristics improve and seed banks are replenished.

The secondary forests created by shifting cultivation are commonly richer in plant and animal resources useful to humans than primary forests, even though they are much less bio-diverse. Shifting cultivators view the forest as an agricultural landscape of fields at various stages in a regular cycle. People unused to living in forests cannot see the fields for the trees. Rather they perceive an apparently chaotic landscape in which trees are cut and burned randomly and so they characterise shifting cultivation as ephemeral or 'pre-agricultural', as 'primitive' and as a stage to be progressed beyond. Shifting agriculture is none of these things. Stable shifting cultivation systems are highly variable, closely adapted to micro-environments and are carefully managed by farmers during both the cropping and fallow stages. Shifting cultivators may possess a highly developed knowledge and understanding of their local environments and of the crops and native plant species they exploit. Complex and highly adaptive land tenure systems sometimes exist under shifting cultivation. Introduced crops for food and as cash have been skillfully integrated into some shifting cultivation systems. Its disadvantages include the high initial cost, as manual labour is required.

Simple Societies and Environmental Change

Shifting cultivation in Indonesia. A new crop is sprouting through the burnt soil.

A growing body of palynological evidence finds that simple human societies brought about extensive changes to their environments before the establishment of any sort of state, feudal or capitalist, and before the development of large scale mining, smelting or shipbuilding industries. In these societies agriculture was the driving force in the economy and shifting cultivation was the most common type of agriculture practiced. By examining the relationships between social and economic change and agricultural change in these societies, insights can be gained on contemporary social and economic change and global environment change, and the place of shifting cultivation in those relationship.

As early as 1930 questions about relationships between the rise and fall of the Mayan civilization of the Yucatán Peninsula and shifting cultivation were raised and continue to be debated today. Archaeological evidence suggests the development of Mayan society and economy began around 250 AD. A mere 700 years later it reached its apogee, by which time the population may have reached 2,000,000 people. There followed a precipitous decline that left the great cities and ceremonial centres vacant and overgrown with jungle vegetation. The causes of this decline are uncertain; but warfare and the exhaustion of agricultural land are commonly cited. More recent work suggests the Maya may have, in suitable places, developed irrigation systems and more intensive agricultural practices.

Similar paths appear to have been followed by Polynesian settlers in New Zealand and the Pacific Islands, who within 500 years of their arrival around 1100 AD turned substantial areas from forest into scrub and fern and in the process caused the elimination of numerous species of birds and animals. In the restricted environments of the Pacific islands, including Fiji and Hawaii, early extensive erosion and change of vegetation is presumed to have been caused by shifting cultivation on slopes. Soils washed from slopes were deposited in valley bottoms as a rich, swampy alluvium. These new environments were then exploited to develop intensive, irrigated fields. The change from shifting cultivation to intensive irrigated fields occurred in association with a rapid growth in population and the development of elaborate and highly stratified chiefdoms. In the larger, temperate latitude, islands of New Zealand the presumed course of events took a different path. There the stimulus for population growth was the hunting of large birds to extinction, during which time forests in drier areas were destroyed by burning, followed the development of intensive agriculture in favorable environments, based mainly on sweet potato (Ipomoea batatas) and a reliance on the gathering of two main wild plant species in less favorable environments. These changes, as in the smaller islands, were accompanied by population growth, the competition for the occupation of the best environments, complexity in social organization, and endemic warfare.

The record of humanly induced changes in environments is longer in New Guinea than in most places. Agricultural activities probably began 5,000 to 9,000 years ago. However, the most spectacular changes, in both societies and environments, are believed to have occurred in the central highlands of the island within the last 1,000 years, in association with the introduction of a crop new to New Guinea, the sweet potato. One of the most striking signals of the relatively recent intensification of agriculture is the sudden increase in sedimentation rates in small lakes.

The root question posed by these and the numerous other examples that could be cited of simple societies that have intensified their agricultural systems in association with increases in population and social complexity is not whether or how shifting cultivation was responsible for the extensive changes to landscapes and environments. Rather it is why simple societies of shifting cultivators in the tropical forest of Yucatán, or the highlands of New Guinea, began to grow in numbers and to develop stratified and sometimes complex social hierarchies.

At first sight, the greatest stimulus to the intensification of a shifting cultivation system is a growth in population. If no other changes occur within the system, for each extra person to be fed from the system, a small extra amount of land must be cultivated. The total amount of land available is the land being presently cropped and all of the land in fallow. If the area occupied by the system is not expanded into previously unused land, then either the cropping period must be extended or the fallow period shortened.

At least two problems exist with the population growth hypothesis. First, population growth in most pre-industrial shifting cultivator societies has been shown to be very low over the long term. Second, no human societies are known where people work only to eat. People engage in social relations with each other and agricultural produce is used in the conduct of these relationships.

These relationships are the focus of two attempts to understand the nexus between human societies and their environments, one an explanation of a particular situation and the other a general exploration of the problem.

Feedback Loops

In a study of the Duna in the Southern Highlands of New Guinea, a group in the process of moving from shifting cultivation into permanent field agriculture post sweet potato, Modjeska argued for the development of two "self amplifying feed back loops" of ecological and social causation. The trigger to the changes were very slow population growth and the slow expansion of agriculture to meet the demands of this growth. This set in motion the first feedback loop, the "use-value" loop. As more forest was cleared there was a decline in wild food resources and protein produced from hunting, which was substituted for by an increase in domestic pig raising. An increase in domestic pigs required a further expansion in agriculture. The greater protein available from the larger number of pigs increased human fertility and survival rates and resulted in faster population growth.

The outcome of the operation of the two loops, one bringing about ecological change and the other social and economic change, is an expanding and intensifying agricultural system, the conversion of forest to grassland, a population growing at an increasing rate and expanding geographically and a society that is increasing in complexity and stratification.

Resources are Cultural Appraisals

The second attempt to explain the relationships between simple agricultural societies and their environments is that of Ellen. Ellen does not attempt to separate use-values from social production. He argues that almost all of the materials required by humans to live (with perhaps the exception of air) are obtained through social relations of production and that these relations proliferate and are modified in numerous ways. The values that humans attribute to items produced from the environment arise out of cultural arrangements and not from the objects themselves, a restatement of Carl Sauer's dictum that "resources are cultural appraisals". Humans frequently translate actual objects into culturally conceived forms, an example being the translation by the Duna of the pig into an item of compensation and redemption. As a result, two fundamental processes underlie the ecology of human social systems: First, the obtaining of materials from the environment and their alteration and circulation through social relations, and second, giving the material a value which will affect how important it is to obtain it, circulate it or alter it. Environmental pressures are thus mediated through social relations.

Transitions in ecological systems and in social systems do not proceed at the same rate. The rate of phylogenetic change is determined mainly by natural selection and partly by human interference and adaptation, such as for example, the domestication of a wild species. Humans however have the ability to learn and to communicate their knowledge to each other and across generations. If most social systems have the tendency to increase in complexity they will, sooner or later, come into conflict with, or into "contradiction" with their environments. What happens around the point of "contradiction" will determine the extent of the environmental degradation that will occur. Of particular importance is the ability of the society to change, to invent or to innovate technologically and sociologically, in order to overcome the "contradiction" without incurring continuing environmental degradation, or social disintegration.

An economic study of what occurs at the points of conflict with specific reference to shifting cultivation is that of Esther Boserup. Boserup argues that low intensity farming, extensive shifting cultivation for example, has lower labor costs than more intensive farming systems. This assertion remains controversial. She also argues that given a choice, a human group will always choose the technique which has the lowest absolute labor cost rather than the highest yield. But at the point of conflict, yields will have become unsatisfactory. Boserup argues, contra Malthus, that rather than population always overwhelming resources, that humans will invent a new agricultural technique or adopt an existing innovation that will boost yields and that is adapted to the new environmental conditions created by the degradation which has occurred already, even though they will pay for the increases in higher labor costs. Examples of such changes are the adoption of new higher yielding crops, the exchanging of a digging stick for a hoe, or a hoe for a plough, or the development of irrigation systems. The controversy over Boserup's proposal is in part over whether intensive systems are more costly in labor terms, and whether humans will bring about change in their agricultural systems before environmental degradation forces them to.

Contemporary Shifting Cultivation Practice

Sumatra, Indonesia.

Rio Xingu, Brazil.

Santa Cruz, Bolivia.

Kasempa, Zambia.

The estimated rate of deforestation in Southeast Asia in 1990 was 34,000 km^2 per year. In Indonesia alone it was estimated 13,100 km^2 per year were being lost, 3,680 km^2 per year from Sumatra and 3,770 km^2 from Kalimantan, of which 1,440 km^2 were due to the fires of 1982 to 1983. Since those estimates were made huge fires have ravaged Indonesian forests during the 1997 to 1998 El Niño associated drought.

Shifting cultivation was assessed by the FAO to be one of the causes of deforestation while logging was not. The apparent discrimination against shifting cultivators caused a confrontation between FAO and environmental groups, who saw the FAO supporting commercial logging interests against the rights of indigenous people. Other independent studies of the problem note that despite lack of government control over forests and the dominance of a political elite in the logging industry, the causes of deforestation are more complex. The loggers have provided paid employment to former subsistence farmers. One of the outcomes of cash incomes has been rapid population growth among indigenous groups of former shifting cultivators that has placed pressure on their traditional long fallow farming systems. Many farmers have taken advantage of the improved road access to urban areas by planting cash crops, such as rubber or pepper. Increased cash incomes often are spent on chain saws, which have enabled larger areas to be cleared for cultivation. Fallow periods have been reduced and cropping periods extended. Serious poverty elsewhere in the country has brought thousands of land-hungry settlers into the cut-over forests along the logging roads. The settlers practice what appears to be shifting cultivation but which is in fact a one-cycle slash and burn followed by continuous cropping, with no intention to long fallow. Clearing of trees and the permanent cultivation of fragile soils in a tropical environment with little attempt to replace lost nutrients may cause rapid degradation of the fragile soils.

The loss of forest in Indonesia, Thailand, and the Philippines during the 1990s was preceded by major ecosystem disruptions in Vietnam, Laos and Cambodia in the 1970s and 1980s caused by warfare. Forests were sprayed with defoliants, thousands of rural forest dwelling people uproots from their homes and moved and roads driven into previously isolated areas. The loss of the tropical forests of Southeast Asia is the particular outcome of the general possible outcomes described by Ellen when small local ecological and social systems become part of larger system. When the previous relatively stable ecological relationships are destabilized, degradation can occur rapidly. Similar descriptions of the loss of forest and destruction of fragile ecosystems could be provided from the Amazon Basin, by large scale state sponsored colonization forest land (Becker 1995, 61) or from the Central Africa where what endemic armed conflict is destabilizing rural settlement and farming communities on a massive scale.

Comparison with Other Ecological Phenomena

In the tropical developing world, shifting cultivation in its many diverse forms, remains a pervasive practice. Shifting cultivation was one of the very first forms of agriculture practiced by humans and its survival into the modern world suggests that it is a flexible and highly adaptive means of production. However, it is also a grossly misunderstood practice. Many casual observers cannot see past the clearing and burning of standing forest and do not perceive often ecologically stable cycles of cropping and fallowing. Nevertheless, shifting cultivation systems are particularly susceptible to rapid increases in population and to economic and social change in the larger world around them. The blame for the destruction of forest resources is often laid on shifting cultivators. But the forces bringing about the rapid loss of tropical forests at the end of the 20th century are the same forces that led to the destruction of the forests of Europe, urbanization, industrialization, increased affluence, populational growth and geographical expansion and the application the latest technology to extract ever more resources from the environment in pursuit of wealth and political power by competing groups. However we must know that those who practice Agriculture are at the receiving end of the social stratum.

Studies of small, isolated and pre-capitalist groups and their relationships with their environments suggests that the roots of the contemporary problem lie deep in human behavioral patterns, for even in these simple societies, competition and conflict can be identified as the main force driving them into contradiction with their environments.

SUBSISTENCE FARMING

Like most farmers in Sub-Saharan Africa, this Cameroonian man cultivates at the subsistence level.

Subsistence farming, or subsistence agriculture, is a mode of agriculture in which a plot of land produces only enough food to feed the family or small community working it. All produce grown is intended for consumption purposes as opposed to market sale or trade. Historically and currently a difficult way of life, subsistence farming is considered by many a backward lifestyle that should be transformed into industrialized communities and commercial farming throughout the world in order to overcome problems of poverty and famine. The numerous obstacles that have prevented this to date suggest that a complex array of factors, not only technological but also economic, political, educational, and social, are involved. An alternative perspective, primarily from the feminist voice, maintains that the subsistence lifestyle holds the key to sustainability as human relationships and harmony with the environment have priority over material measures of wealth. Although the poverty suffered by many of those who have never developed beyond subsistence levels of production in farming is something that needs to be overcome, it does appear that the ideas inherent in much of subsistence farming—cooperation, local, ecologically appropriate—are positive attributes that must be preserved in our efforts to improve the lives of all people throughout the world.

Subsistence farming is a mode of agriculture in which a plot of land produces only enough food to feed those who work it—little or nothing is produced for sale or trade. Depending on climate, soil conditions, agricultural practices and the crops grown, it generally requires between 1,000 and 40,000 square meters (0.25 to 10 acres) per person.

A recognizably harsh way of living, subsistence farmers can experience a rare surplus of produce goods under conditions of good weather which may allow farmers to sell or trade such goods at market. Because such surpluses are rare, subsistence farming does not allow for consistent economic growth and development, the accumulation of capital, or the specialization of labor. Diets of subsistence communities are confined to little else than what is produced by community farmers. Subsistence crops are usually organic due to a lack of finances to buy or trade for industrial inputs such as fertilizer, pesticides or genetically modified seeds.

Techniques

In the absence of technology, the area of land that a farmer can cultivate each season is limited by factors such as available tools and the quality of the soil. Tools used by subsistence farmers are often primitive. Most farmers do not have access to large domesticated work animals, and therefore clear, toil, and harvest their goods using pointed sticks, hoes, or by hand.

Techniques of subsistence farming include "slash and burn" clearing in which farmers clear plots of farmland by cutting down all brush, allowing the debris to dry, and later burning the fallen refuse. This works to clear the field for cultivation, while the leftover ash serves as a natural fertilizer. This type of clearing technique is often employed by subtropical communities throughout lush areas of South and Central America, and parts of Indonesia.

If the land does not produce a surplus, due to the fertility of the soil, climate conditions, tools and techniques, or available crop types, the farmer can do no more than hope to subsist on it. Under these conditions, subsequent years with poor harvests often result in food scarcity and famine.

Not all subsistence farmers have access to as much land as they can cultivate. Many times, socioeconomic conditions prevent an expansion of farming plots and any increase in produce levels. If inheritance traditions require that a plot be split among an owner's children upon the owner's death, plot sizes steadily decrease.

Subsistence Farming and the Modern World

Despite its difficulties, subsistence farming remains a part of the modern world today. For many underdeveloped nations, subsistence farming represents the only option to prevent starvation and famine.

Subsistence farming in Zambia.

Subsistence farming has been argued to be economically efficient within various subtropical regions of Columbia and Papua New Guinea. Under these subtropical conditions, rainfall levels are often high and various crops can be produced year round. Due to these conditions, production levels often prove adequate enough to provide for small subsistence farming communities.

This argument does not hold for many Sub-Saharan regions of Africa, where poverty and famine levels are some of the highest in the world. One reason why subsistence farming systems have failed throughout the Sub-Saharan region are increasing trends in population growth that are not met with an equal increase in the production of agricultural output. Other reasons include unusually harsh climate conditions, widespread disease among plants and animals, and a lack of efficient institutional structures.

In parts of rural Zambia, much of the current population relies on subsistence farming to survive. As irrigation systems are few, most Zambians must rely on seasonal rains to ensure crop production. In 1995, Zambia underwent a severe drought which vastly diminished production levels throughout traditional farming communities. Similar impoverishment has been observed throughout parts of the Amazon Basin of Brazil and the Indonesian islands of Sumatra and Borneo, which also rely heavily on subsistence farming and production.

Many developmental economists have argued against the use of subsistence farming and instead promote commercial farming and economic industrialization as the solution to worldwide hunger. Economist Ronald E. Seavoy, author of Subsistence and Economic Development, argued that subsistence farming is to blame for high levels of poverty and increasing instances of famine, recommending the transformation of subsistence agriculture into commercial agriculture which would ultimately promote economic development among economically underdeveloped nations.

Attempts have continued to be made to move in this direction away from subsistence farming. In central Uganda commercial farming has been promoted to alleviate high poverty levels throughout Ugandan subsistence farming communities. Restructuring the production output of the people and identifying a potential market for free trade, are key to successful small-scale industrialization, thereby improving rural living conditions and diminishing poverty rates.

An alternative viewpoint, particularly promoted by women often called "ecofeminists," reflects the need to understand sustainable economies. Those such as Maria Mies and Vandana Shiva have argued that the free market capitalist system is inherently unsustainable in the long run, since it exploits various population groups and the environment. Instead, they argue that the "catch-up" model of economic development, assuming that western-style progress is possible and optimal for all, be replaced with a more ecologically sensitive approach, valuing harmony with nature and the goals of happiness, quality of life, and human dignity over the accumulation of wealth. They explain subsistence as empowerment for all, based on people's strengths and their cooperation with nature and each other.

A similar proposal was made in the early twentieth century by the American First Lady and "First-wave" feminist Eleanor Roosevelt, speaking in favor of "subsistence homesteads" within the United States.

"The objective of subsistence farming is not to compete with regular farming or add to the burden of agricultural overproduction. The idea is that families engaged in subsistence farming consume their own garden products locally instead of sending them to distant markets. They are not expected to support themselves entirely by raising food, like the successful commercial farmers of the country. The plan is that they shall be situated near enough to an industry for one member of the family to be employed in a factory a sufficient number of days in the year to bring in the amount of money needed to pay for the things which the families must have and cannot produce for themselves. In this way farming will be helped by industry, and industry will be helped by farming."

Subsistence farming as practiced in many parts of the world in the past and today is not the ideal model. Nonetheless, subsistence farming generally recognizes the cycles of renewability in nature and works in harmony with them to maintain productivity. Ultimately, sustainability depends upon harmony among people and with our environment. There are lessons to be learned in this regard from the subsistence perspective.

MIXED CROPPING

Mixed cropping, also known as polyculture, inter-cropping, or co-cultivation, is a type of agriculture that involves planting two or more plants simultaneously in the same field, interdigitating the crops so that they grow together. In general, the theory is that planting multiple crops at once saves space since crops in the same field might ripen at different seasons, and provides a wealth of environmental benefits. Documented benefits of mixed cropping include the balance of input and outgo of soil nutrients, the suppression of weeds and insect pests, the resistance of climate extremes (wet, dry, hot, cold), the suppression of plant diseases, the increase in overall productivity, and the management of scarce resources (land) to the fullest degree.

Planting enormous fields with single crops is called monocultural agriculture, and it is a recent invention of the industrial agricultural complex. Most agricultural field systems of the past involved some form of mixed cropping, although unambiguous archaeological evidence of this is difficult to come by. Even if botanical evidence of the plant residues (such as starches or phytoliths) of multiple crops is discovered within an ancient field, it has proven difficult to differentiate between the results of mixed cropping and rotation cropping. Both methods are believed to have been used in the past.

The primary reason for prehistoric multi-cropping probably had more to do with the needs of the farmer's family, rather than any recognition that mixed cropping was a good idea. It is possible that certain plants adapted to multi-cropping over time, as a result of the domestication process.

Classic Mixed Cropping

The classic example of mixed cropping is that of the American "three sisters": maize, beans, and cucurbits (squash and pumpkins).

The three sisters were domesticated at different times but eventually were combined together to form an important component of Native American agriculture and cuisine. The mixed cropping of the three sisters is historically documented by the Seneca and Iroquois tribes in the US northeast and probably began sometime after 1000 C.E. The method consists of planting all three seeds in the same hole. As they grow, the maize provides a stalk for the beans to climb on, the beans are nutrient-rich to offset that taken out by the maize, and the squash grows low to the ground to keep weeds down and keep water from evaporating from the soil in the heat.

Modern Mixed Cropping

Agronomists studying mixed crops have had mixed results determining if yield differences can be achieved with mixed versus monoculture crops. For example, a combination wheat and chickpeas might work in one part of the world, but it might not work in another. But, overall it appears that measurably good effects result when the right combination of crops are cropped together.

Mixed cropping is best suited for small-scale farming where harvesting is by hand. It has been used to improve income and food production for small farmers and lessen the likelihood of total crop failure—even if one of the crops fail, the same field might still produce other crop successes. Mixed cropping also requires fewer nutrient inputs such as fertilizers, pruning, pest control, and irrigation than does monoculture farming.

Benefits

There seems to be no doubt that the practice provides a rich biodiverse environment, fostering habitat and species richness for animals and insects such as butterflies and bees. Some evidence suggests that polycultural fields do produce high yields compared to monocultural fields in certain situations, and almost always increase biomass richness over time. Polyculture in forests, heathlands, grasslands, and marshes has been particularly important for the regrowth of biodiversity in Europe.

A recent study (Pech-Hoil and colleagues) was conducted on the tropical American perennial achiote (Bixa orellana), a rapidly growing tree that has a high carotenoid content, and a food dye and spice in small farming cultures in Mexico. The experiment looked at achiote as it is grown in different agronomic systems—intercropped polyculture, backyard cultivation including poultry farming, and a wide range of plants, and monoculture. Achiote adapted its mating system depending on which type of system it was planted in, specifically the amount of outcrossing that is seen.

INTERCROPPING

Intercropping is a multiple cropping practice involving growing two or more crops in proximity. The most common goal of intercropping is to produce a greater yield on a given piece of land by making use of resources or ecological processes that would otherwise not be utilized by a single crop.

Potential Benefits

Resource Partitioning

Careful planning is required, taking into account the soil, climate, crops, and varieties. It is particularly important not to have crops competing with each other for physical space, nutrients, water, or sunlight. Examples of intercropping strategies are planting a deep-rooted crop with a shallow-rooted crop, or planting a tall crop with a shorter crop that requires partial shade. Inga alley cropping has been proposed as an alternative to the ecological destruction of slash-and-burn farming.

When crops are carefully selected, other agronomic benefits are also achieved.

Mutualism

Planting two crops in close proximity can especially be beneficial when the two plants interact in a way that increases one or both of the plant's fitness (and therefore yield). For example, plants that are prone to tip over in wind or heavy rain (lodging-prone plants), may be given structural support by their companion crop. Climbing plants can also benefit from structural support. Some plants are used to suppress weeds or provide nutrients. Delicate or light-sensitive plants may be given shade or protection, or otherwise wasted space can be utilized. An example is the tropical multi-tier system where coconut occupies the upper tier, banana the middle tier, and pineapple, ginger, or leguminous fodder, medicinal or aromatic plants occupy the lowest tier.

Intercropping of compatible plants can also encourage biodiversity, by providing a habitat for a variety of insects and soil organisms that would not be present in a single-crop environment. These organisms may provide crops valuable nutrients, such as through nitrogen fixation.

Pest Management

There are several ways in which increasing crop diversity may help improve pest management. For example, such practices may limit outbreaks of crop pests by increasing predator biodiversity. Additionally, reducing the homogeneity of the crop can potentially increase the barriers against biological dispersal of pest organisms through the crop.

There are several ways pests can be controlled through intercropping:

- Trap cropping, this involves planting a crop nearby that is more attractive for pests compared to the production crop, the pests will target this crop and not the production crop.
- Repellant intercrops, an intercrop that has a repellent effect to certain pests can be used. This system involved the repellant crop masking the smell of the production crop in order to keep pests away from it.
- Push-pull cropping, this is a mixture of trap cropping and repellant intercropping. An attractant crop attracts the pest and a repellant crop is also used to repel the pest away.

The degree of spatial and temporal overlap in the two crops can vary somewhat, but both requirements must be met for a cropping system to be an intercrop. Numerous types of intercropping, all of which vary the temporal and spatial mixture to some degree, have been identified. These are some of the more significant types:

- Mixed intercropping, as the name implies, is the most basic form in which the component crops are totally mixed in the available space.
- Row cropping involves the component crops arranged in alternate rows. Variations include alley cropping, where crops are grown in between rows of trees, and strip cropping, where multiple rows, or a strip, of one crop are alternated with multiple rows of another crop. A new version of this is to intercrop rows of solar photovoltaic modules with agriculture crops. This practice is called agrivoltaics.
- Temporal intercropping uses the practice of sowing a fast-growing crop with a slow-growing crop, so that the fast-growing crop is harvested before the slow-growing crop starts to mature.
- Further temporal separation is found in relay cropping, where the second crop is sown during the growth, often near the onset of reproductive development or fruiting, of the first crop, so that the first crop is harvested to make room for the full development of the second.

Limitations

Intercropping to reduce pest damage in agriculture, has been deployed with varying success. For example, while trap cropping has reduced pest densities at a commercial in experiments, it often fails to decrease pest densities deployed in large scale commercial landscapes. Furthermore, increasing crop diversity through intercropping does not necessarily increase the presence of the predators of crop pests. In a systematic review of the literature, in 2008, in the studies examined, predators of pests tended only increased under crop diversification strategies in 53 percent of studies, and crop diversification only led to increased yield in only 32% of the studies.

Chili pepper intercropped with coffee in Colombia's southwestern Cauca Department.

Coconut and Mexican marigold, a multilayer cropping in India.

MONOCROPPING

Monocropping is the agricultural practice of growing a single crop year after year on the same land, in the absence of rotation through other crops or growing multiple crops on the same land polyculture. Corn, soybeans, and wheat are three common crops often grown using monocropping techniques.

Monocropping allows for farmers to have consistent crops throughout their entire farm. They can use the same seed, pest control, machinery and growing method on their entire farm giving them a larger yield for a significantly lower cost.

Soil Ecology

While economically a very efficient system, allowing for specialization in equipment and crop production, monocropping is also controversial, as it damages the soil ecology (including depletion or reduction in diversity of soil nutrients) and provide an unbuffered niche for parasitic species, increasing crop vulnerability to opportunistic insects, plants, and microorganisms. The result is a more fragile ecosystem with an increased dependency on pesticides and artificial fertilizers. The concentrated presence of a single cultivar, genetically adapted with a single resistance strategy, presents a situation in which an entire crop can be wiped out very quickly by a single opportunistic species. An example of this would be the potato famine of Ireland in 1845–1849, and according to Devlin Kuyek is the main cause of the current food crisis with monoculture rice crops failing as the effects of climate change become more acute.

Strategy

Monocropping as an agricultural strategy tends to emphasize the use of expensive specialized farm equipment—an important component in realizing its efficiency goals. This can lead to an increased dependency on fossil fuels and reliance on expensive machinery that cannot be produced locally and may need to be financed. This can make a significant change in the economics of farming in regions that are accustomed to self-sufficiency in agricultural production. In addition, political complications may ensue when these dependencies extend across national boundaries.

The controversies surrounding monocropping are complex, but traditionally the core issues concern the balance between its advantages in increasing short-term food production—especially in hunger-prone regions—and its disadvantages with respect to long-term land stewardship and the fostering of local economic independence and ecological sustainability. Advocates of monocropping tend to claim that in its absence many human populations would be reduced to starvation or to a degraded level of civilization comparable to the Dark Ages. On the other hand, critics of monocropping dispute these claims and attribute them to corporate special interest groups, citing the damage that monocropping causes to societies and the environment.

Monocropping Difficulty

A difficulty with monocropping is that the solution to one problem—whether economic, environmental or political—may result in a cascade of other problems. For example, a well-known concern is pesticides and fertilizers seeping into surrounding soil and groundwater from extensive monocropped acreage in the U.S. and abroad. This issue, especially with respect to the pesticide DDT, played an important role in focusing public attention on ecology and pollution issues during the 1960s when Rachel Carson published her landmark book *Silent Spring*.

Soil depletion is also a negative effect of mono-cropping. Crop rotation plays an important role in replenishing soil nutrients, especially atmospheric nitrogen converted to usable forms by nitrogen-fixing plants used in fallow fields. In addition, it performs an important role in preventing pathogen and pest build-up. In a monocropping regime, farmers are less likely to rotate their crops and replenish such essential soil nutrients. In addition, artificial high-nitrogen fertilizers can "burn" the soil by creating an unfavorable environment for indigenous organisms, a phenomenon well-known to organic gardeners and farmers (who avoid it), resulting in further disruption of soil ecology and dependence on further short-term fertilizer strategies. Lacking a stable ecology, in the absence of substantial irrigation and chemical "fixes", the soil can become dry and begin to erode. As the soil becomes arid and useless, the need for more land becomes an issue, leading to the destruction of even more land—a high-tech version of slash and burn agriculture.

Deforestation

Under certain circumstances monocropping can lead to deforestation or the displacement of indigenous peoples. For example, since 1970 the Amazon Rainforest has lost nearly 1/5 of its forest cover. A main cause of this deforestation is local farmers clearing land for more crops. In Colombia, the need for more farming land is causing the displacement of large populations of peasants.

Subsidies

In order to help reduce dependence on fossil fuels the U.S. government subsidizes the monocropping of corn and soybeans to be used in ethanol production. However, monocropping itself is highly chemical- and energy-intensive, as studies by Nelson (2006) indicate. Such studies have shown that the "hidden" energy costs associated with producing each unit of bio-fuel are significantly larger than the amount of energy available from the fuel itself.

COMPANION PLANTING

Companion planting agriculture is the planting of different crops in proximity for any of a number of different reasons, including pest control, pollination, providing habitat for beneficial insects, maximizing use of space, and to otherwise increase crop productivity. Companion planting is a form of polyculture.

Companion planting is used by farmers and gardeners in both industrialized and developing countries for many reasons. Many of the modern principles of companion planting were present many centuries ago in cottage gardens in England and forest gardens in Asia, and thousands of years ago in Mesoamerica.

Mechanisms

Companion planting can operate through a variety of mechanisms, which may sometimes be combined.

Provision of Nutrients

Legumes such as clover provide nitrogen compounds to other plants such as grasses by fixing nitrogen from the air with symbiotic bacteria in their root nodules.

Dandelions have long taproots that bring nutrients from deep within the soil to near the surface, benefitting neighboring plants that are shallower-rooted.

Trap Cropping

Trap cropping uses alternative plants to attract pests away from a main crop. For example, nasturtium (*Tropaeolum majus*) is a food plant of some caterpillars which feed primarily on members of the cabbage family (brassicas); some gardeners claim that planting them around brassicas protects the food crops from damage, as eggs of the pests are preferentially laid on the nasturtium. However, while many trap crops have successfully diverted pests off of focal crops in small scale greenhouse, garden and field experiments, only a small portion of these plants have been shown to reduce pest damage at larger commercial scales.

Host-Finding Disruption

Recent studies on host-plant finding have shown that flying pests are far less successful if their host-plants are surrounded by any other plant or even "decoy-plants" made of green plastic, cardboard, or any other green material.

The host-plant finding process occurs in phases:

- The first phase is stimulation by odours characteristic to the host-plant. This induces the insect to try to land on the plant it seeks. But insects avoid landing on brown (bare) soil. So if only the host-plant is present, the insects will quasi-systematically find it by simply landing on the only green thing around. This is called (from the point of view of the insect) "appropriate landing". When it does an "inappropriate landing", it flies off to any other nearby patch of green. It eventually leaves the area if there are too many 'inappropriate' landings.
- The second phase of host-plant finding is for the insect to make short flights from leaf to leaf to assess the plant's overall suitability. The number of leaf-to-leaf flights varies according to the insect species and to the host-plant stimulus received from each leaf. The insect must accumulate sufficient stimuli from the host-plant to lay eggs; so it must make a certain number of consecutive 'appropriate' landings. Hence if it makes an 'inappropriate landing', the assessment of that plant is negative, and the insect must start the process a new.

Thus it was shown that clover used as a ground cover had the same disruptive effect on eight pest species from four different insect orders. An experiment showed that 36% of cabbage root flies laid eggs beside cabbages growing in bare soil (which resulted in no crop), compared to only 7% beside cabbages growing in clover (which allowed a good crop). Simple decoys made of green cardboard also disrupted appropriate landings just as well as did the live ground cover.

Pest Suppression

Some companion plants help prevent pest insects or pathogenic fungi from damaging the crop, through chemical means. For example, the smell of the foliage of marigolds is claimed to deter aphids from feeding on neighbouring plants.

Predator Recruitment

Companion plants that produce copious nectar or pollen in a vegetable garden (insectary plants) may help encourage higher populations of beneficial insects that control pests, as some beneficial predatory insects only consume pests in their larval form and are nectar or pollen feeders in their adult form. For instance, marigolds with simple flowers attract nectar-feeding adult hoverflies, the larvae of which are predators of aphids.

Protective Shelter

Shade-grown coffee plantation in Costa Rica. The red trees in the background provide shade; those in the foreground have been pruned to allow full exposure to the sun.

Some crops are grown under the protective shelter of different kinds of plant, whether as wind breaks or for shade. For example, shade-grown coffee, especially Coffea arabica, has traditionally been grown in light shade created by scattered trees with a thin canopy, allowing light through to the coffee bushes but protecting them from overheating. Suitable Asian trees include Erythrina subumbrans (tton tong or dadap), Gliricidia sepium (khae falang), Cassia siamea (khi lek), Melia azedarach (khao dao sang), and Paulownia tomentosa, a useful timber tree.

MONOCULTURE

Monoculture is the agricultural practice of producing or growing a single crop, plant, or livestock species, variety, or breed in a field or farming system at a time. Polyculture, where more than one crop is grown in the same space at the same time, is the alternative to monoculture. Monoculture is widely used in both industrial farming and organic farming and has allowed increased efficiency in planting and harvest while simultaneously increasing the risk of exposure to diseases or pests.

Continuous monoculture, or monocropping, where the same species is grown year after year, can lead to the quicker buildup of pests and diseases, and then rapid spread where a uniform crop is susceptible to a pathogen. Monocultures of African palm oil, sugar cane, pines, and soybeans can all be particularly aggressive to environment. The practice has been criticized for its environmental effects and for having possible long term effects on agriculture and food supplies. Diversity can be added both in time, as with a crop rotation or sequence, or in space, with a polyculture.

Oligoculture has been suggested to describe a crop rotation of just a few crops, as is practiced by several regions of the world.

Agriculture

The term is used in agriculture and describes the practice of planting the same cultivar over an extended area. Examples of monoculture include lawns and most fields of wheat or corn. The term is also used where a single breed of farm animal is raised in large-scale concentrated animal feeding operations (CAFOs). In the United States, The Livestock Conservancy was formed to protect nearly 200 endangered livestock breeds from going extinct, largely due to the increased reliance on just a handful of highly specialized breeds.

Diversity of crops in space and time; monocultures and polycultures, and rotations of both.					
			Diversity in time		
			Low	Higher	
				Cyclic	Dynamic (non-cyclic)
Diversity in space	Low	Monoculture, one species in a field	Continuous monoculture, monocropping	Crop rotation (rotation of monocultures)	Sequence of monocultures
	Higher	Polyculture, two or more species intermingled in a field	Continuous polyculture	Rotation of polycultures	Sequence of polycultures

Benefits

In crop monocultures, each cultivar has the same standardized planting, maintenance and harvesting requirements resulting in greater yields and lower costs. For example, researchers have discovered a native plant to Senegal, called Guiera senegalensis, grown next to millet increased millet production roughly 900 percent. When a crop is matched to its well-managed environment, a monoculture can produce higher yields than a polyculture. In the last 40 years, modern practices such as monoculture planting and the use of synthesized fertilizers have reduced the amount of additional land needed to produce food.

Risks

Annually planting the same crop in the same area depletes the nutrients from the earth that the plant relies on and leaves soil weak and unable to support healthy growth. Because soil structure and quality is so poor, farmers are forced to use chemical fertilizers to encourage plant growth and fruit production. These fertilizers, in turn, disrupt the natural makeup of the soil and contribute further to nutrient depletion. Monocropping also creates the spread of pests and diseases, which have to be treated with yet more chemicals. The effects of monocropping on the environment are severe when pesticides and fertilizers make their way into ground water or become airborne, creating pollution. Polyculture, the mixing of different crops, reduces the likelihood that one or more of the crops will be resistant to any particular pathogen. Studies have shown that planting a mixture of crop strains in the same field can combat disease effectively. Switching to polyculture in areas with disease conditions can greatly increased yields. In one study in China, the planting of several varieties of rice in the same field increased yields of non-resistant strains by 89% compared to non-resistant strains grown in monoculture, largely because of a dramatic (94%) decrease in the incidence of disease, making pesticides less necessary.

Humans rely heavily on a relatively small number of food crops and farm animals for food. If disease hits a major food crop - as happened during the 19th-century Irish potato famine - food supplies for large populations could come under threat. Maintaining and increasing biodiversity in agriculture could help safeguard world food-supplies.

Genetic Monocultures

While often referring to the mass production of the same species of crop, it can also refer to planting of a single cultivar which has same identical genetic makeups to the plants around them. When all plants in a monoculture are genetically similar, a disease, to which they have no resistance, can destroy entire populations of crops. As of 2009 the wheat leaf-rust fungus occasioned a great deal of worry internationally, having already decimated wheat crops in Uganda and Kenya, and having started to make inroads into Asia as well. Given the very genetically similar strains of much of the world's wheat crops following the Green Revolution, the impacts of such diseases threaten agricultural production worldwide.

Irish Potato Famine

In Ireland, exclusive use of one variety of potato, the "lumper", led to the Great Famine of 1845-1849. Lumpers provided inexpensive food to feed the Irish masses. Potatoes were propagated vegetatively with little to no genetic variation. When *Phytophthora infestans* arrived in Ireland from the Americas in 1845, the lumper had no resistance to the disease, leading to the nearly complete failure of the potato crop across Ireland.

Bananas

Until the 1950's, the Gros Michel cultivar of banana represented almost all bananas consumed in the United States because of their taste, small seeds, and efficiency to produce. Their small seeds, while more appealing than the large ones in other Asian cultivars, were not suitable for planting. This meant that all new banana plants had to be grown from the cut suckers of another plant. As a result of this asexual form of planting, all bananas grown had identical genetic makeups which gave them no traits for resistance to *Fusarium wilt*, a fungal disease that spread quickly throughout the Caribbean where they were being grown. By the beginning of the 1960's, growers had to switch to growing the Cavendish banana, a cultivar grown in a similar way. This cultivar is under similar disease stress since all the bananas are clones of each other and could easily succumb as the Gros Michel did.

Cattle

Many of today's livestock production systems rely on just a handful of highly specialized breeds. Focusing heavily on a single trait (output) may come at the expense of other desirable traits - such as fertility, resistance to disease, vigor, and mothering instincts. In the early 1990s a few Holstein calves were observed to grow poorly and died in the first 6 months of life. They were all found to be homozygous for a mutation in the gene that caused Bovine Leukocyte Adhesion Deficiency. This mutation was found at a high frequency in Holstein populations worldwide. (15% among bulls in the US, 10% in Germany, and 16% in Japan.) Researchers studying the pedigrees of affected and carrier animals tracked the source of the mutation to a single bull that was widely used in the industry. Note that in 1990 there were approximately 4 million Holstein cattle in the US, making the affected population around 600,000 animals.

Benefits of Genetic Diversity

While having little to no variety in the genetics of an agricultural system can have drawbacks, increasing genetic diversity by introducing organisms with varying genes can divert them and make the system more sustainable. For example, by having crops with varying genetic traits for disease and pest resistance, there is a much lower chance of having those pests or diseases spread throughout the area. This is because if one crop becomes infected with a particular strain of disease or species of pest, there is a chance that the other plants around it will have genes that protect them from that strain/species. This can help increase crop productivity while simultaneously lowering pesticide usage and risk of exposure.

POLYCULTURE

Polyculture is a form of agriculture in which more than one species is grown at the same time and place in imitation of the diversity of natural ecosystems. Polyculture is the opposite of monoculture, in which only members of one plant or animal species are cultivated together. Polyculture has traditionally been the most prevalent form of agriculture in most parts of the world and is growing in popularity today due to its environmental and health benefits. There are many types of polyculture including annual polycultures such as intercropping and cover cropping, permaculture, and integrated aquaculture. Polyculture is advantageous because of its ability to control pests, weeds, and disease without major chemical inputs. As such, polyculture is considered a sustainable form of agriculture. However, issues with crop yield and biological competition have caused many modern major industrial food producers to continue to rely on monoculture instead.

Common Practices

The kinds of plants that are grown, their spatial distribution, and the time that they spend growing together determines the specific type of polyculture that is implemented. There is no limit for the types of plants or animals that can be grown together to form a polyculture. The time overlaps between plants can be asymmetrical as well, with one plant depending on the other for longer than is reciprocated, often due to differences in life spans.

Annual Polycultures

Intercropping

When more than two crops are grown in complete spatial and temporal overlap with each other, it is referred to as intercropping. Intercropping is particularly useful in plots with limited land availability. Legumes are one of the most commonly intercropped crops, specifically legume-cereal mixtures. Legumes fix atmospheric nitrogen into the soil so that it is available for consumption by other plants in a process known as nitrogen fixation. The presence of legumes consequently eliminates the need for man-made nitrogen fertilizers in intercrops.

Cover Cropping

When a crop is grown alongside another plant that is not a crop, the combination is referred to as cover cropping. If the non-crop plant is a weed, the combination is called a weedy culture. Grasses and legumes are the most common cover crops. Cover crops are greatly beneficial as they can help prevent soil erosion, physically suppress weeds, improve surface water retention, and, in the case of legumes, provide nitrogen compounds as well.

Strip Cropping

Strip cropping is a form of polyculture that involves growing different plants in alternating rows. While strip cropping does not involve the complete intermixing of plant species, it still provides many of the same benefits such as preventing soil erosion and aiding with nutrient cycling.

Permaculture

A coffee farm in Colombia where coffee plants are grown under several tree species in imitation of natural ecosystems. Trees provide valuable resources for the coffee plants such as shade, nutrients, and soil structure.

Permacultures are polycultures of perennial plants. Legume-grass mixtures and wildflower mixtures are common forms of permaculture, popular in Europe and more temperate climates. Permacultures increase soil fertility through nitrogen fixation, decrease soil erosion, regulate water consumption, and decrease the need for tillage thereby conserving soil nutrients. Permacultures require even less human intervention than other forms of polyculture because of lower harvest and tillage rates.

In many Latin American countries, agroforestry is a popular form of permaculture as well where trees and crops are grown together. Trees provide shade for the crops alongside organic matter and nutrients when they shed their leaves or fruits. The elaborate root systems of trees also help prevent soil erosion and increase the presence of microbes in the soil. In addition to benefiting crops, trees act as commodities themselves for use in paper, medicine, firewood, etc. Growing coffee plants alongside other tree species in Mexico is a common practice of agroforestry.

Coffee is a shade-loving crop, and is traditionally shade-grown. In India, it is often grown under a natural forest canopy, replacing the shrub layer. A different polyculture system is used for coffee in Mexico, where the *Coffea* bushes are grown under leguminous trees in the genus *Inga*.

Integrated Aquaculture

Integrated aquaculture is a form of aquaculture in which cultures of fish or shrimp are grown together with seaweed, shellfish, or micro-algae. Mono-species aquaculture, a form of aquaculture where only members of the species are grown together, poses several problems for farmers and the environment. The harvesting of seaweed crops in monocultures, for example, releases nitrates into the water and can lead to severe eutrophication as has occurred in the Venice Lagoon. In terms of seafood growth, the greatest problem in monocultures is the high cost of feed, which accounts for about half of production costs. However, more than half of seafood feed is shown to go to waste and can lead to further problems with excess nitrogen release and eutrophication or algal blooms of freshwater. Many technological approaches to lowering these harmful environmental effects such as bacterial bio-filters have proved to consume high levels of energy and be economically costly.

As such, many farmers have transitioned towards integrated aquaculture. In integrated aquaculture farms, plants serve a dual purpose, acting as food for the sea animals and as a water filtration device for the surrounding environment by absorbing nitrates and excess oxygen. Nutrients can be recycled between plants and animals, reducing the need for chemical nutrient supplements. Plants that are grown alongside seafood such as seaweed often hold significant commercial value by themselves, so incorporating them into already existing seafood monocultures increases economic value.

Functions

Pest Management

Pests are less predominant in polycultures than monocultures due to crop diversity. The lack of concentration of a single crop makes polycultures less appealing to pests who have a strong preference towards a specific crop. These specialized pests will often have more difficulty locating a favorable host plant inside of a polyculture than in a monoculture. If a pest has more generalized preferences, it will leave more quickly to other plants in the polyculture and as such have a lesser effect on any one plant. When pests are present in the nearby area, polycultures consequently experience lower yield loss than monocultures do in a theory known as the associational resistance hypothesis. Because polycultures mimic naturally diverse ecosystems, general pests are less likely to distinguish between polycultures and the surrounding environment as well. As such, pests travel more freely between the two environments, and have a relative smaller in presence in polycultures to begin with.

Because of the diversity of plants in a polyculture, natural enemies, or predators, of pests are often attracted to the polyculture alongside pests themselves as well. These natural enemies help suppress pest populations while doing no harm to the plants themselves.

Disease Control

Plant diseases are less predominant in polycultures than monocultures. The disease-diversity hypothesis states that a greater diversity of plants leads to a decreased severity of disease. Because different plants are susceptible to different diseases, if a disease negatively impacts one crop, it will not necessarily spread to another and so the overall impact is contained. However, the type of disease and the susceptibility of the specific plants inside the polyculture to a particular disease can vary greatly.

Weed Management

Both the density of crops and the diversity affect weed growth in polycultures. Having a greater density of plants reduces the available water, sunlight, and nutrient concentrations in the environment. Such a reduction is heightened with greater crop diversity as more potential resources are fully utilized. This level of competition makes polycultures particularly inhospitable to weeds.

When they do grow, weeds can help polycultures, assisting in pest management by attracting natural enemies of pests. They can also act as hosts to arthropods that are beneficial to other plants in the polyculture.

Advantages

Sustainability

Applying pesticides to crops in a monoculture: Some of these pesticides will likely end up in water sources and the atmosphere where they can pose serious health and environmental harm. Polycultures attract fewer pests than monocultures and experience less yield loss even when pests are present, often eliminating the need for pesticides.

Because polycultures use methods of pest, disease, and weed control that do not rely on human intervention, they do not release pesticides into the environment. Fertilizer use is reduced as well, as diverse plants more fully share and use all available soil and atmospheric nutrients. As such, environmental impacts such as eutrophication of fresh water or the presence of excess atmospheric nitrogen are greatly reduced.

Other negative impacts of modern agriculture are similarly reduced. Excessive tillage occurs in most modern agricultural practices, but removes essential microbes and nutrients from the soil that are conserved in polycultures, especially permacultures. Because polyculture relies on natural systems of crop maintenance, farmers save money on machinery. By growing multiple plants or animals together in the same space, agricultural land, a critical resource worldwide, taking up 40% of the world's land area, is used more productively.

Polyculture increases local biodiversity. Increasing crop diversity can increase pollination in nearby environments, as diverse plants attract a broader array of pollinators. This is one example of reconciliation ecology, or accommodating biodiversity within human landscapes. This may also form part of a biological pest control program.

Human Health

The chemicals used in monoculture food production can be directly harmful to human health when released into the environment. Nitrogen is a chemical found in especially high concentrations in fertilizers. Nitrates from these fertilizers often become integrated in water sources due to agricultural runoff. The consumption of nitrates at high doses has been shown to lead to methemoglobinemia in infants.

Many of the crops consumed today are calorie-rich crops as well which can lead to illnesses such as obesity, hypertension, and type II diabetes. Because it encourages plant diversity, polyculture can help increase diet diversity by incorporating non-traditional foods into agriculture and people's diets.

Effectiveness

The effects of interspecific competition and intraspecific competition can cause great damage to plants in certain polycultures. In order for a polyculture to be effective, the diverse species that are a part of it must have distinct biological needs such as absorbing different nutrients or requiring different amounts of sunlight as stated by the competitive exclusion principle. Due to the large number of plant species that are cultivated by humans, finding and testing combinations of plants where interspecific and intraspecific competition do not significantly negatively affect the individual plants is extremely difficult. As such, for crops where historic polycultures do not exist, such a multiplicity makes the creation of new polycultures a significant issue.

Crop yield is also an issue in polycultures. While a polyculture produces more biomass overall than a monoculture, individual crops inside of the polyculture are not as prevalent. When there is a focal crop whose cultivation is especially important for a society a lower yield for a certain crop may pose food availability issues.

Similarly, while diseases and pests affect a polyculture less as a group, they do not necessarily have a decreased effect on a focal crop. If targeted by a specialized pest or disease, a focal crop in a polyculture will likely experience the same yield loss as its monoculture counterpart. Polyculture also often requires more labor.

PLANTATION FARMING

Plantation farming is a form of commercial farming where crops are grown for the purpose of making profits. The fact that crops are grown for profits means that large tracts of land are needed to make this type of farming successful. This form of farming is mainly found in areas that experience tropical climate. Research has shown that many people don't know the difference between plantation farming and other types of farming. In fact, most people usually think that they are the same.

- The yields produced is for commercial purpose: One of the main difference between plantation farming and other types of farming is that the main goal of plantation farming is that crop are grown for commercial purposes. That is not usually the case with other types of farming. In most cases, other types of farming such as subsistence farming usually produce yields that can be used for personal or small commercial use. However, that is not the case with plantation farming. The main aim of plantation agriculture is to produce crops that will be used for commercial purposes. Its aim to serve large market.

- Only one type of crop is grown: Another main difference between plantation farming and other types of farming is that only one type of crop is grown. For instance, if the owner decides to plant wheat, the whole plantation will have wheat. Other crops will not be mixed. This strategy is usually used to enhance specialization and productivity. It also help to reduce the cost of production.

- Heavy capital outlay: Plantation farming is grown for commercial purposes. As a result, the capital is needed in order to make this projects success is high. This form of farming is labor intensive and highly mechanized. As a results, they require high capital outlay because cost of production is also high. However, that is not the case with other types of farming which require low capital outlays.

- Farms are highly mechanized and scientifically managed: Plantation farming require large tracts of land. Managing huge tracts of land is not easy. It require efficient management in order to be successful. As a results, this type of farming is highly mechanized and scientifically managed. This helps to boost efficiency and production. However, that is not the case with other types of farming. In fact some types of farming such as subsistence farming usually use simple tools to cultivate.

- Farms are large: Plantation farming can only be successful if it is cultivated in large tracts of land. This is because this type of farming is meant to sustain huge population. Plantation farming requires more than 40 hectares in order to be successful. Other types of farming does not require huge tracts of land in order to be successful. Other types of farming can be done regardless of the size of the land available.

FACTORS AFFECTING CROP PRODUCTION

Internal Factors Affecting Crop Production

Genetic Factors

The increase in crop yields and other desirable characters are related to Genetic make up of plants:

- High yielding ability,
- Early maturity,
- Resistance to lodging,
- Drought flood and salinity tolerance,
- Tolerance to insect pests and diseases,
- Chemical composition of grains (oil content, protein content),
- Quality of grains (fineness, coarseness),
- Quality of straw (sweetness, juiciness).

The above characters are less influenced by environmental factors since they are governed by genetic make-up of crop.

External Factors Affecting Crop Production

- Climatic,
- Edaphic,
- Biotic,
- Phsiographic,
- Socio-economic.

Climatic Factors

Nearly 50 % of yield is attributed to the influence of climatic factors.The following are the atmospheric weather variables which influences the crop production.

- Precipitation,
- Temperature,
- Atmospheric humidity,
- Solar radiation,
- Wind velocity,
- Atmospheric gases.

Precipitation

- Precipitation includes all water which falls from atmosphere such as rainfall, snow, hail, fog and dew.
- Rainfall one of the most important factor influences the vegetation of a place.
- Total precipitation in amount and distribution greatly affects the choice of a cultivated species in a place.
- In heavy and evenly distributed rainfall areas, crops like rice in plains and tea, coffee and rubber in Western Ghats are grown.
- Low and uneven distribution of rainfall is common in dryland farming where drought resistance crops like pearl millet, sorghum and minor millets are grown.
- In desert areas grasses and shrubs are common where hot desert climate exists.
- Though the rainfall has major influence on yield of crops, yields are not always directly proportional to the amount of Precipitation as excess above optimum reduces the yields.
- Distribution of rainfall is more important than total rainfall to have longer growing period especially in drylands.

Temperature

- Temperature is a measure of intensity of heat energy. The range of temperature for maximum growth of most of the agricultural plants is between 15 and 40 ^{o}C.
- The temperature of a place is largely determined by its distance from the equator (latitude) and altitude.

- It influences distribution of crop plants and vegetation.
- Germination, growth and development of crops are highly influenced by temperature.
- Affects leaf production, expansion and flowering.
- Physical and chemical processes within the plants are governed by air temperature.
- Diffusion rates of gases and liquids changes with temperature.
- Solubility of different substances in plant is dependent on temperature.
- The minimum, maximum (above which crop growth ceases) and optimum temperature of individual's plant is called as cardinal temperature.

Crops	Minimum Temperature °C	Optimum Temperature °C	Maximum Temperature °C
Rice	10	32	36-38
wheat	4-5	20	30-32
Maize	8-10	20	40-43
Sorghum	12-13	25	40
Tobacco	12-14	29	35

Atmospheric Humidity (Relative Humidity - RH)

- Water is present in the atmosphere in the form of invisible water vapour, normally known as humidity. Relative humidity is ratio between the amount of moisture present in the air to the saturation capacity of the air at a particular temperature.
- If relative humidity is 100% it means that the entire space is filled with water and there is no soil evaporation and plant transpiration.
- Relative humidity influences the water requirement of crops.
- Relative humidity of 40-60% is suitable for most of the crop plants.
- Very few crops can perform well when relative humidity is 80% and above.
- When relative humidity is high there is chance for the outbreak of pest and disease.

Solar Radiation (Without which Life will not Exist)

- From germination to harvest and even post harvest crops are affected by solar radiation.

- Biomass production by photosynthetic processes requires light.
- All physical process taking place in the soil, plant and environment are dependent on light.
- Solar radiation controls distribution of temperature and there by distribution of crops in a region.
- Visible radiation is very important in photosynthetic mechanism of plants. Photosynthetically Active Radiation (PAR - 0.4 – 0.7μ) is essential for production of carbohydrates and ultimately biomass.
    - 0.4 to 0.5 μ - Blue - violet - Active,
    - 0.5 to 0.6 μ - Orange - red - Active,
    - 0.5 to 0.6 μ - Green - yellow - Low active.
- Photoperiodism is a response of plant to day length Short day – Day length is 12 hours (Barley, oat, carrot and cabbage), day neutral – There is no or less influence on day length (Tomato and maize).
- Phototropism – Response of plants to light direction. Eg. Sunflower.
- Photosensitive – Season bound varieties depends on quantity of light received.
- Wind velocity.
- The basic function of wind is to carry moisture (precipitation) and heat.
- The moving wind not only supplies moisture and heat, also supplies fresh CO_2 for the photosynthesis.
- Wind movement for 4 – 6 km/hour is suitable for more crops.
- When wind speed is enormous then there is mechanical damage of the crops (i.e.) it removes leaves and twigs and damages crops like banana, sugarcane.
- Wind dispersal of pollen and seeds is natural and necessary for certain crops.
- Causes soil erosion.
- Helps in cleaning produce to farmers.
- Increases evaporation.
- Spread of pest and diseases.

Atmospheric Gases on Plant Growth

- CO_2 – 0.03%, O_2 - 20.95%, N_2 - 78.09%, Argon - 0.93%, Others - 0.02%.

- CO_2 is important for Photosynthesis, CO_2 taken by the plants by diffusion process from leaves through stomata.
- CO_2 is returned to atmosphere during decomposition of organic materials, all farm wastes and by respiration.
- O_2 is important for respiration of both plants and animals while it is released by plants during Photosynthesis.
- Nitrogen is one of the important major plant nutrient, Atmospheric N is fixed in the soil by lightning, rainfall and N fixing microbes in pulses crops and available to plants.
- Certain gases like SO_2, CO, CH_4, HF released to atmosphere are toxic to plants.

Edaphic Factors (Soil)

Plants grown in land completely depend on soil on which they grow. The soil factors that affect crop growth are:

- Soil moisture
- Soil air
- Soil temperature
- Soil mineral matter
- Soil organic matter
- Soil organisms
- Soil reactions

Soil Moisture

- Water is a principal constituent of growing plant which it extracts from soil.
- Water is essential for photosynthesis.
- The moisture range between field capacity and permanent wilting point is available to plants.
- Available moisture will be more in clay soil than sandy soil.
- Soil water helps in chemical and biological activities of soil including mineralization.
- It influences the soil environment Eg. it moderates the soil temperature from extremes.

- Nutrient availability and mobility increases with increase in soil moisture content.

Soil Air

- Aeration of soil is absolutely essential for the absorption of water by roots.
- Germination is inhibited in the absence of oxygen.
- O_2 is required for respiration of roots and micro organisms.
- Soil air is essential for nutrient availability of the soil by breaking down insoluble mineral to soluble salts.
- For proper decomposition of organic matter.
- Potato, tobacco, cotton linseed, tea and legumes need higher O_2 in soil air.
- Rice requires low level of O_2 and can tolerate water logged (absence of O_2) condition.

Soil Temperature

- It affects the physical and chemical processes going on in the soil.
- It influences the rate of absorption of water and solutes (nutrients).
- It affects the germination of seeds and growth rate of underground portions of the crops like tapioca, sweet potato.
- Soil temperature controls the microbial activity and processes involved in the nutrient availability.
- Cold soils are not conducive for rapid growth of most of agricultural crops.

Soil Mineral Matter

- The mineral content of soil is derived from the weathering of rocks and minerals as particles of different sizes.
- These are the sources of plant nutrients.
- E.g.; Ca, Mg, S, Mn, Fe, K etc.

Soil Organic Matter

- It supplies all the major, minor and micro nutrients to crops.
- It improves the texture of the soil.

- It increases the water holding capacity of the soil.
- It is a source of food for most microorganisms.
- Organic acids released during decomposition of organic matter enables mineralization process thus releasing unavailable plant nutrients.

Soil Organisms

The raw organic matter in the soil is decomposed by different micro organisms which in turn releases the plant nutrients. Atmospheric nitrogen is fixed by microbes in the soil and is available to crop plants through symbiotic (Rhizobium) or non-symbiotic (Azospirillum) association.

Soil Reaction (PH)

- Soil reaction is the pH (hydrogen ion concentration) of the soil.
- Soil pH affects crop growth and neutral soils with pH 7.0 are best for growth of most of the crops.
- Soils may be acidic (<7.0), neutral (=7.0), saline and alkaline (>7.0).
- Soils with low pH is injurious to plants due high toxicity of Fe and Al.
- Low pH also interferes with availability of other plant nutrients.

Biotic Factors

Beneficial and harmful effects caused by other biological organism (plants and animals) on the crop plants.

Plants

- Competitive and complimentary nature among field crops when grown together.
- Competition between plants occurs when there is demand for nutrients, moisture and sunlight particularly when they are in short supply or when plants are closely spaced.
- When different crops of cereals and legumes are grown together, mutual benefit results in higher yield (synergistic effect).
- Competition between weed and crop plants as parasites eg: Striga parasite weed on sugarcane crop.

Animals

- Soil fauna like protozoa, nematode, snails, and insects help in organic matter decomposition, while using organic matter for their living.

- Insects and nematodes cause damage to crop yield and considered as harmful organisms.
- Honey bees and wasps help in cross pollination and increases yield and considered as beneficial organisms.
- Burrowing earthworm facilitates aeration and drainage of the soil as ingestion of organic and mineral matter by earthworm results in constant mixing of these materials in the soils.
- Large animals cause damage to crop plants by grazing (cattle, goats etc.).

Physiographic Factors

- Topography is the nature of surface earth (leveled or sloppy) is known as topography. Topographic factors affect the crop growth indirectly.
- Altitude: Increase in altitude cause a decrease in temperature and increase in precipitation and wind velocity (hills and plains).
- Steepness of slope: It results in run off of rain water and loss of nutrient rich top soil.
- Exposure to light and wind: A mountain slope exposed to low intensity of light and strong dry winds may results in poor crop yields (coastal areas and interior pockets).

Socio-Economic Factors

- Society inclination to farming and members available for cultivation.
- Appropriate choice of crops by human beings to satisfy the food and fodder requirement of farm household.
- Breeding varieties by human invention for increased yield or pest & disease resistance.
- The economic condition of the farmers greatly decides the input/resource mobilizing ability (marginal, small, medium and large farmers).

References

- Types-crop-production-system: jliedu.com, Retrieved 10 July, 2019
- Miguel Angel Altieri; Clara Ines Nicholls (2004). Biodiversity and Pest Management in Agroecosystems, Second Edition. Psychology Press. ISBN 9781560229230
- "Controlling Pests with Plants: The power of intercropping - UVM Food Feed". UVM Food Feed. 2014-01-09. Retrieved 2016-12-01

- Subsistence-farming, entry: newworldencyclopedia.org, Retrieved 29 June, 2019
- "Cucurbitaceae--Fruits for Peons, Pilgrims, and Pharaohs". University of California at Los Angeles. Archived from the original on October 16, 2013. Retrieved September 2, 2013
- Mixed-cropping-history-171201: thoughtco.com, Retrieved 30 January, 2019
- "Pacific Northwest Nursery IPM. Flowers, Sweets and a Nice Place to Stay: Courting Beneficials to Your Nursery". Oregon State University. Retrieved 11 February 2013
- What-is-plantation-farming:driftwoodplantation.com, Retrieved 31 March, 2019
- Haspel, Tamar (9 May 2014). "Monocrops: They're a problem, but farmers aren't the ones who can solve it". New York Times. Retrieved 12 January2016

Crop Diseases

CHAPTER 4

Crops are affected by numerous diseases such as botrytis cinerea, Phomopsis stem, pod blight, common scab, rust disease, Ascochyta blight and bacterial blight. A few of the soil based crop diseases are rhizoctonia, verticillium and crown rot. All these crop diseases have been carefully analyzed in this chapter.

Crop disease involves any harmful deviation or alteration from the normal functioning of the physiological processes. Therefore, diseased plants suffer disturbances from normal life processes and their vital functions.

Cabbage infected by bacterial blight disease.

In an attempt to reach high yields and healthy crops, farmers throughout the world struggle to prevent and eradicate various diseases from their crops. Each crop is susceptible to particular diseases that affect the quality and final yield potential.

Generally, it's estimated that various pests (insects, weeds, nematodes, animals, diseases) each year cause crop yield losses of 20-40%. More precisely, there is some data that maintains that crop diseases cause average yield losses of 42% for the most important food crops. In some cases, crop diseases destroy the whole crop production.

For this reason, it's extremely important for farmers to find out all they can about the crop diseases so they can manage them properly.

What Causes Crop Diseases?

The disease usually occurs and spreads from season to season, depending on the presence of a certain pathogen, as well as environmental conditions and the characteristics of each crop variety. Essentially, crop diseases occur according to the nature of their causal agent:

- Abiotic or Noninfectious disease agents.
- Biotic or Infectious disease agents.

Marginal leaves necrosis caused by freezing temperatures.

Abiotic, or Noninfectious disease agents, include non-living environmental conditions or inappropriate farm management. They are not transmissable to other plants. A few universally recognized abiotic agents are:

- Extreme temperatures,
- Moisture,
- Wind,
- Frequent and heavy rain,
- Drought or flood,
- Excess or deficiency of nutrients,
- Soil compaction,
- Chemical injury caused by pesticides or salts,
- Improper water management.

Biotic, or Infectious disease agents, are living organism pathogens capable of spreading from one host to another and transmitting the disease.

Black or stem rust caused by the fungus Puccina graminis tritici.

The pathogens are classified as:

- Fungi; the most common pathogens, cause around 85% of plant diseases; examples include Black or stem rust caused by the fungus Puccina graminis tritici,
- Viruses; are transmitted by a vector or attack the plant through a wound; for example, the Apple mosaic virus affects apple, plum, and hazelnut,
- Bacteria; mutate and multiply rapidly; they enter the plant through a wound or stomata; for instance, the Apple fire blight is caused by Erwinia amylovora,
- Nematodes; damage the crops causing galls on roots,
- Parasitic plants, live on crops, as they lack chlorophyll they obtain it from the host plant; for example, the dwarf mistletoe grows on other plants and derives nutrients from the host,
- Algae; theoretically don't cause significant damage, however, the may under special circumstances.

Apple fire blight.

Apple mosaic virus.

How Disease occurs in the Plant

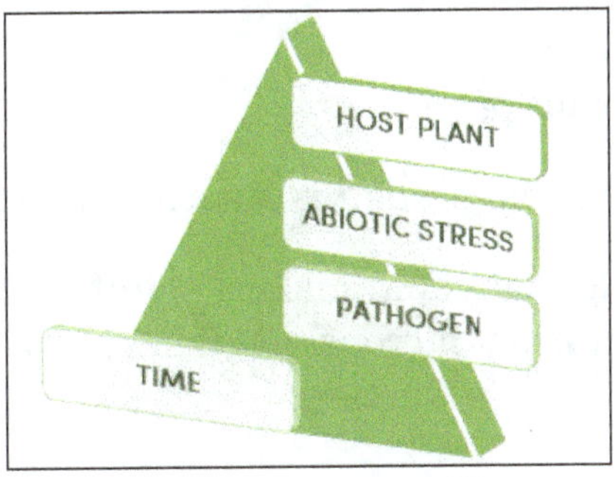

Plant disease pyramid.

Understanding the disease and the development process is the first step towards successful disease management. There are a few special conditions that are conducive for disease development. First, each crop is susceptible to some disease. Second, abiotic factors, namely weather, weaken the plant significantly. Therefore, the plant is vulnerable to the attack of the pathogen. So, the pathogen acts like a cherry on top. When all of the aforementioned factors (susceptible crop, abiotic stress, pathogen attack) are present together and at the same time, disease occurs. That is called the Plant Disease Pyramid.

How to Save the Yields and Quality of Crops?

Saving the yields and protecting the crops from diseases is a burning issue for every farmer who wishes to see the best results. From the beginning, the first and crucial step in crop protection is knowledge. Every farmer should be aware of susceptibility of the crop to certain diseases, as well as the abiotic agents that are favorable for the occurrence of the disease.

More importantly, one of the most powerful practices is practicing preventative measures, such as:

- Planting resistant varieties.
- Managing optimal planting and harvesting times.
- Planting healthy and quality material.
- Dezinfestation of equipment.
- Crop rotation.
- Managing plant nutrients according to crop needs.

Outside the preventative measures, there are also a few curative measures that should be taken in the case of the disease actually occurring. These measures include:

- The using of various chemicals, including fungicides, nematicides, bactericides, algacides.
- Removing infested plant parts.
- Utilizing various beneficial predators such as viruses, fungi, or bacteria.

There is one additional thing that could make a significant difference in disease farm management. Regular monitoring of crops and fields, as well as on-time reaction, can be a true lifesaver for every crop production.

BOTRYTIS CINEREA

Botrytis cinerea is a necrotrophic fungus that affects many plant species, although its most notable hosts may be wine grapes. In viticulture, it is commonly known as "botrytis bunch rot"; in horticulture, it is usually called "grey mould" or "gray mold".

The fungus gives rise to two different kinds of infections on grapes. The first, grey rot, is the result of consistently wet or humid conditions, and typically results in the loss of the affected bunches.

The second, noble rot, occurs when drier conditions follow wetter, and can result in distinctive sweet dessert wines, such as Sauternes. Although poetic, the "grapes" refers to the bunching of the fungal spores on their conidiophores, and "ashes" just refers to the greyish colour of the spores *en masse*. The fungus is usually referred to by its anamorph (asexual form) name, because the sexual phase is rarely observed. The teleomorph (sexual form) is an ascomycete, *Botryotinia fuckeliana*, also known as *Botryotinia cinerea*.

Hosts and Symptoms

Hosts

The disease, gray mold, affects more than 200 dicotyledonous plant species and a few monocotyledonous plants found in temperate and subtropical regions. Serious economic losses can be a result of this disease to both field and greenhouse grown crops. The causal agent, Botrytis cinerea can infect mature or senescent tissues, plants prior to harvest, or seedlings. There is a wide variety of hosts infected by this pathogen including protein crops, fiber crops, oil crops, and horticultural crops. Horticultural crops include vegetables (examples are chickpeas, lettuce, broccoli, and beans) and small fruit crops (examples are grape, strawberry, and raspberry), these are most severely affected and devastated by gray mold. Plant organs affected include fruits, flowers, leaves, storage organs, and shoots.

Symptoms and Signs

Symptoms vary across plant organs and tissues. *B. cinerea* is a soft rot that will have a collapsed and water soaked appearance on soft fruit and leaves. Brown lesions may develop slowly on undeveloped fruit. Twigs infected with gray mold will die back. Blossoms will cause fruit drop and injury, such as ridging on developing and mature fruit. Symptoms are visible at wound sites where the fungus begins to rot the plant. Gray masses with a velvety appearance are conidia on the plant tissues are a sign of plant pathogen. These conidia are asexual spores that will continue to infect the plant and surrounding hosts throughout the growing season making this a polycyclic disease.

Plants have evolved to produce localized lesions when a pathogen attacks. An oxidative burst causes hypersensitive cell death called a hypersensitive response (HR). This soft rot can trigger HR to assist in colonization. *Botrytis cinerea*, as a necrotrophic pathogen, exploits the dead tissue for its pathogenicity or its ability to cause disease. Susceptible plants cannot use the HR to protect against *Botrytis cinerea*.

Biology

A Botrytis cinerea conidiophore.

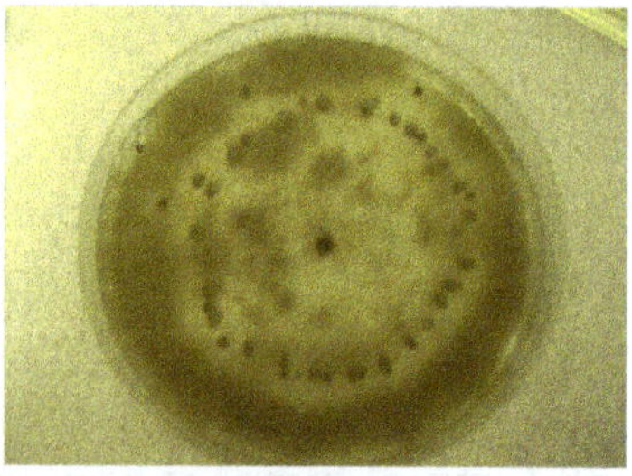

Botrytis cinerea growing on a plate with a ring of visible sclerotia.

Botrytis cinerea is characterized by abundant hyaline conida (asexual spores) borne on grey, branching tree-like conidiophores. The fungus also produces highly resistant sclerotia as survival structures in older cultures. It overwinters as sclerotia or intact mycelia, both of which germinate in spring to produce conidiophores. The conidia, dispersed by wind and by rain-water, cause new infections.

Different Botrytis cinerea strains show considerable genetic variability (polyploidy).

Gliocladium roseum is a fungal parasite of Botrytis cinerea.

Environment

Gray mold favors moist, humid, and warm environmental conditions between 18.3-23-9 °C (65-75 °F). Temperature, relative humidity, and wetness duration produce a conducive environment that is favorable for inoculation of mycelium or conidia. Controlled environments, such as crop production greenhouses, provide the moisture and high temperatures that favor the spreading and development of the pathogen *Botrytis cinerea.*

Standing water on plant leaf surfaces provides a place for spores to germinate. Humid conditions can result from improper irrigation practice, plants placed too close together, or the structure of the greenhouse not allowing for efficient ventilation and air flow. Ventilation at night significantly reduces the incidence of gray mold.

Melanized sclerotium allows *Botrytis cinerea* to survive for years in the soil. Sclerotia and the asexual conidia spores contribute to the widespread infection of the pathogen.

A low pH is preferred by the gray mold to perform well. *Botrytis cinerea* can acidify its environment by secreting organic acids, like oxalic acid. By acidifying its surroundings, cell wall degrading enzymes (CWDEs) are enhanced, plant-protection enzymes are inhibited, stomatal closure is deregulated, and pH signaling is mediated to facilitate its pathogenesis.

Viticulture

In the *Botrytis* infection known as "noble rot" (*pourriture noble* in French, or *Edelfäule* in German), the fungus removes water from the grapes, leaving behind a higher percent of solids, such as sugars, fruit acids and minerals. This results in a more intense, concentrated final product. The wine is often said to have an aroma of honeysuckle and a bitter finish on the palate.

A distinct fermentation process initially caused by nature, the combination of geology, climate and specific weather led to the particular balance of beneficial fungus while leaving enough of the grape intact for harvesting. The Chateau d'Yquem is the only Premier Cru Supérieur, largely due to the vineyard's susceptibility to noble rot.

Botrytis complicates winemaking by making fermentation more complex. *Botrytis* produces an anti-fungal that kills yeast and often results in fermentation stopping before the wine has accumulated sufficient levels of alcohol. Makers of fine German dessert wines have been known to take fermenting tubs of wine into their homes to nurture the yeast through the night to assure that the alcohol level reaches legal minimums for the product to be called wine.

Botrytis cinerea on Riesling grapes.

Botrytis bunch rot is another condition of grapes caused by *Botrytis cinerea* that causes great losses for the wine industry. It is always present on the fruitset, however, it requires a wound to start a bunch rot infection. Wounds can come from insects, wind, accidental damage, etc. To control botrytis bunch rot there are a number of fungicides available on the market. Generally, these should be applied at bloom, bunch closure and veraison (the most important being the bloom application). Some winemakers are known to use the German method of fermentation and prefer having a 5% bunch rot rate in their grapes and will usually hold the grapes on the vine a week longer than normal.

Horticulture

Botrytis cinerea affects many other plants. It is economically important on soft fruits such as strawberries and bulb crops. Unlike wine grapes, the affected strawberries are not edible and are discarded. To minimize infection in strawberry fields, good ventilation around the berries is important to prevent moisture being trapped among leaves and berries. A number of bacteria have been proven to act as natural antagonists to B. cinerea in controlled studies.

In greenhouse horticulture, Botrytis cinerea is well known as a cause of considerable damage in tomatoes.

The infection also affects rhubarb, snowdrops, white meadowfoam, western hemlock, Douglas-fir and cannabis. Potassium bicarbonate-based fungicide has been proven to cure and prevent powdery mildew, blackspot, downy mildew, blights, molds and other plant diseases, such as Botrytis cinerea.

Human Disease

Botrytis cinerea mold on grapes may cause "winegrower's lung", a rare form of hypersensitivity pneumonitis (a respiratory allergic reaction in predisposed individuals).

Mycoviruses of Botrytis Cinerea

Botrytis cinerea not only infects plants, it also hosts several mycoviruses itself.

Table: Mycoviruses of Botrytis cinerea.

Genus	Species	Appearance	Genome	Phenotypical change
Unassigned	Unassigned	Isometric particles, 30-45 nm; bacilliform, 25x63 nm	dsRNA not determined	None, latent infection
Unassigned	Unassigned	Naked dsRNA	dsRNA not determined	None, latent infection
Unassigned	Botrytis cineria virus- CVg25	Isometric VLP's(L dsRNA), ~40 nm	dsRNA (L, M_1, and M_2), 8.3 kb	None, latent infection
Mycoflexivirus	Botrytis virus F	Flexuous, 270 nm	+ssRNA, ~6.8 kb	None, latent infection
Unassigned	Unassigned	Isometric, 33 nm	+ssRNA, ~6.8 kb	Reduced sporulation, lac case activity, and invasivity
Botrexvirus	Botrytis virus X	Flexuous, 270 nm	+ssRNA, ~6.9 kb	None, latent infection
Unassigned	Unassigned (BcMV1)	Naked dsRNA	dsRNA, 3.0 kb	Abnormal mitochondria, reduced mycelia growth

A range of phenotypic alterations due to the mycoviral infection have been observed from symptomless to mild impact, or more severe phenotypic changes including reduction in pathogenicity, growth/suppression of mycelia, sporulation and sclerotia production, formation of abnormal colony sectors and virulence.

Management

Botrytis cinerea can be managed through cultural, chemical, and biological practices.

There are no resistant species to the gray mold rot. Gray mold can be culturally controlled by monitoring the amount and timing of fertilizer applications to reduce the amount of fruit rot. Excessive application of nitrogen will increase the incidence of disease while not improving yields.

Planting cultivars that don't have an upright or dense growth habit can reduce disease as these limit airflow and are favorable for the pathogen. Spacing of plants so they are not touching will increase airflow allowing the area to dry out and reduce the spread of disease. Pruning or purposeful removal of diseased, dead, or overgrown limbs on a regular schedule can also help to improve air movement.

Sanitation by removing dead or dying plant tissue in the fall will decrease inoculum levels as there is no debris for the sclerotium or mycelia to overwinter. Removing debris in the spring will remove inoculum from the site. Disposal of berries during harvest that have signs and symptoms of gray mold will reduce inoculum for the following year.

Biochar, a form of charcoal, can be applied as a soil amendment to strawberry plants to reduce the severity of the fungal disease by stimulating defense pathways within the plant.

Gray mold can be chemically controlled with well-timed fungicide applications starting during the first bloom. Timing can reduce the chance of resistance and will save on costs.

Biological controls or microbial antagonists used for disease suppression, have been successfully used in Europe and Brazil in the form of fungi-like Trichoderma harzianium Rifai and Gliocladium roseum Bainier. Trichoderma species especially, have been shown to control gray mold.

PHOMOPSIS STEM AND POD BLIGHT OF LUPINS

This disease is caused by the fungus Diaporthe toxica. Infection can cause minor crop losses, however, the major impact of infection is the production of a mycotoxin (phomopsin) in mature lupin stems. The toxin can cause sickness or death (lupinosis) of stock that graze infected lupin stubble or seed.

The fungus can infect stems, leaves, pods and seed of lupins. Prematurely dying plants after pod set can be seen in crops and are found particularly in parts of the paddock that are stressed by drought, frost or herbicides.

Symptoms usually appear on senescing lupin stems and are the appearance of dark purplish lesions that bleach with age and contain black fruiting bodies and can cause plants to lodge, The appearance of disease symptoms while plants are green is uncommon. Infected seedlings develop deep-yellow to brown, irregular shaped lesions on their stems below the cotyledons. Severe lesions may girdle the stem and kill the plant. Sometimes, yellow to brown blotches are visible on the leaves and stems of older plants, but most infected plants show no disease symptoms until late in the season at senescence. The surface of green pods can become 'slimy' and mature pods may be shrivelled with reddish-tan to black lesions.

Pod lesions can lead to infected seeds which appear as either normal or are discoloured light yellow to reddish-tan. Web-like grey mould of the fungus may be seen on the seed coat and inside the seed pod.

The disease will most likely be seen on lupin stubble. Small black fruiting bodies develop, often after summer rain which stimulates growth of the fungus and the production of toxins, Once this occurs care should be taken when grazing stubbles for risk of livestock poisoning.

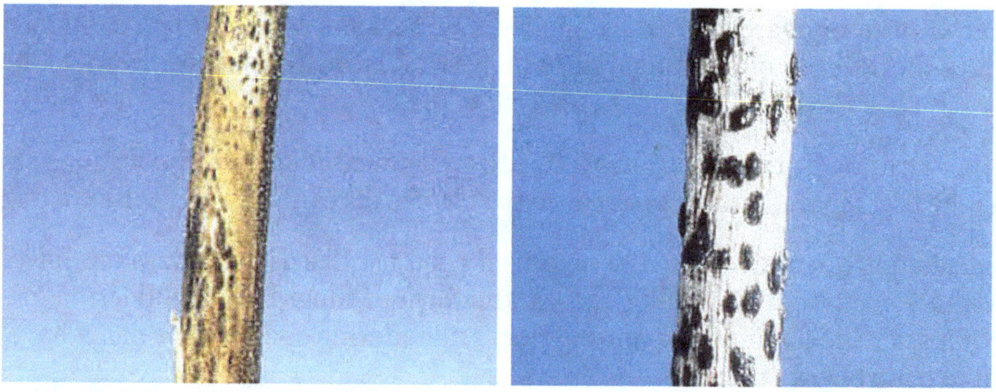

Black fungal fruiting bodies are embedded in infected lupin trash after harvest.

Disease Cycle

The fungus can survive on lupin trash and seed. On trash, the fungus can remain viable for several years. The disease usually becomes established when spores produced on old lupin stubble are carried by wind and rain into new crops. Infection can occur at any stage of plant growth if conditions are favourable for the disease.

Sowing infected seed may give rise to infected seedlings. Sometimes these develop severe stem lesions and die shortly after emergence. Plants that do survive often lack vigour and may lodge. Occasionally spore producing lesions form on infected plants while they are still green, often following a period of moisture stress.

Management

Prolonged rainfall or high humidity in late spring and summer can favour the pathogen and some toxicity may still develop requiring care to be taken when grazing occurs.

Paddock Selection

Crop rotation is an important management strategy and increasing the break between lupin crops allows weathering and breakdown of infected stubble, reducing disease inoculum. Don't sow lupins in the vicinity of old infected lupin stubbles. Infected crop residues should be destroyed by deep ploughing or burning.

Resistant Varieties

In areas where lupinosis has been reported, only Phomopsis resistant lupin varieties should be grown. Variety selection is an important management strategy and an effective way of reducing phomopsis. Current varieties have moderate levels of resistance to both stem and pod infection.

Seed Treatment: Use disease-free seed.

RUST DISEASES OF WHEAT

Rust diseases represent the most economically significant fungal diseases in wheat and other cereal crops worldwide, and they are widely distributed across wheat growing regions. With the capacity to form new strains of fungus, rusts can attack even previously resistant varieties. Rust diseases possess the ability to spread and travel long distances by dispersal of windborne spores and can rapidly develop under optimal weather conditions.

The three rust diseases affecting wheat are leaf, stem and stripe rust. Leaf rust is the most common of the three diseases in the Central Great Plains and other wheat-growing regions in the United States. In some states, leaf rust disease occurs every year. Stem rust is not typically as prevalent as other rusts because many varieties are now resistant to the disease. Stripe rust is becoming an increasingly important disease, with recent outbreaks in the Great Plains states.

Identification and Lifecycle

Stripe rust, also known as yellow rust, is caused by the fungus Puccinia striiformis f. sp. tritici. The disease primarily occurs on leaves, although glumes and awns may also be affected. This part of the lifecycle leads to "hot spots" of infection seen in crops in later winter and early spring.

In lower humidity, stripe rust spores disperse more freely into the air and can travel for much greater distances. This may result in a uniform pattern of disease development beginning in mid-spring.

In recent years, stripe rust has become more common in the Great Plains states. It reached high levels in early spring 2015 due an unusually wet spring across most of the region.

Stripe rust appears earlier in the season because development is enhanced by the cool, moist weather early in the growing season, versus leaf rust, which is more prevalent later in the spring when temperatures warm. Stripe rust develops under ideal temperatures of 45 to 54 degrees F with six to eight hours of leaf moisture.

Stripe rust typically produces yellow or orange blister-like pustules that are arranged in stripes. The lesions produce massive amounts of spores that are easily dislodged and dispersed by wind. The spores may leave orange dust on the clothing of individuals walking through heavily infected fields.

When stripe rust spores land on a living wheat leaf, germination and infection can result. Growing inside the leaf, stripe rust produces new lesions containing new spores. Under high-humidity conditions during winter, most spores survive in small clumps that are relatively heavy and fall quickly out of the air when dispersed. As a result, spores spread mostly over short distances and may appear in uniform patterns of disease seen in mid-spring. Several cycles of spore production occur throughout the growing season.

Crop Damage

It's possible for stripe rust to cause 100 percent crop loss in susceptible varieties if the disease begins early in the season under wet, cool weather conditions.

Managing Wheat Rusts

Farmers have several options to manage wheat rusts, ranging from seed treatments, variety selection, scouting, cultural practices and fungicide applications.

Seed Treatments

Using the correct fungicide seed treatment and rate is one way to provide effective and economical disease control in wheat production. Seed treatments protect seed and young plants from disease and other threats to plant health and yield. Areas commonly infested with rust also will need a foliar fungicide treatment to provide protection beyond the seedling growth stage.

Variety Selection

Where available, use rust-resistant varieties for best protection against leaf rusts. Every commercially available wheat variety has a unique disease package, and excellent disease resistance is not available to manage all disease threats in high-yielding varieties. It's best to select two or three high-yielding varieties that offer the best resistance to common diseases found on your individual farm.

Scouting

When scouting for weeds and insects, check for the presence of wheat rusts and other diseases. Monitor reports of wheat rust development occurring in states south of your area. This will allow you to track the progression of rust diseases migrating north from overwintering hosts and will also help you predict the timing and severity of infestations before they might affect your region. Keep a close eye on weather conditions because rust spores spread through wind currents to promote disease infection.

If you suspect leaf rust, stem rust or stripe rust infection, take samples and work with your county Extension agent to confirm a diagnosis. Crop identification guides are also helpful. Scouting helps determine levels of infection so you can make the best decision about the necessity and rate of fungicide applications.

Cultural Practices

Disease-free seed gives seedlings a good start. Good weed control preplant, at planting and throughout the growing season also helps protect against disease and other pests.

Fungicides

Good control of wheat rusts can be achieved with commercially available fungicides and proper application timing. The decision to use fungicides should be based on scouting for symptoms. It's important to assess disease severity from the onset of infection through the various growth stages. Application timing should take into consideration that diseases should be managed before infection reaches the upper leaves. Other factors affecting fungicide application are infection levels in the field, the susceptibility of the variety and the market price for wheat grain.

Crop Science Solutions to Prevent Wheat Rusts

A well-thought-out disease-management program, including best management practices, proper seed protection and selection and fungicide applications using multiple modes of action, should be implemented to sustainably manage diseases. The following Crop Science solutions are valuable tools to consider for your program.

Because cereal fungal diseases can overwinter and survive in crop residue, crop rotations that include dicot crops can help reduce the inoculum of fungal leaf spot pathogens. Fungicide seed treatments can provide a healthy start for seedlings, especially in cool and damp spring conditions. EverGol Energy from Crop Science is a seed treatment fungicide that promotes more root growth for faster crop establishment and controls seed and soil borne disease such as Rhizoctonia. It features a new combination of fungicides incorporating a complementary mode of action that supports resistance management.

A number of fungicides are available for both early-season and late-season control of these common leaf diseases. Chemistries from two of the most commonly used classes of fungicides – triazoles and strobilurins – provide good to excellent activity against wheat leaf diseases and, when used in conjunction with best management practices, can help manage disease resistance. Wheat growers should consider fungicides with systemic movement and curative properties for the broadest protection from cereal foliar diseases.

With a combination of two chemistries, Prosaro fungicide provides preventive and curative action against key cereal leaf diseases such as various types of rust, Septoria leaf blotch, tan spot and powdery mildew. Additionally, Prosaro provides unsurpassed activity against head diseases such as scab (Fusarium head blight) and glume blotch. It's a good choice to ensure grain quality and enhance yield potential.

BOTRYTIS GREY MOULD OF LENTIL

Botrytis grey mould of lentils is a foliar disease caused by the fungal pathogens Botrytis cinerea and Botrytis fabae. Botrytis grey mould (BGM) has been reported in many lentil growing countries of Asia and America but appears to be more severe in Australia.

The disease can infect all above ground parts of the lentil plant, including leaves, stems and pods. The first symptoms are greyish-brown lesions on the lower leaves of the plant. However, no black fruiting bodies are produced within the lesions.

Characteristic stem infection caused by botrytis. Note grey, fuzzy spore mass and black sclerotia.

Infected leaves can become covered with a fuzzy layer of grey mould near the base of the plant. The stem lesions will appear light brown or bleached and can be covered in grey, mouldy growth. The mouldy growth consists of spore masses that can quickly spread to infect surrounding plants. Clouds of spores can rise from infected plants if they are disturbed. Dark black fungal bodies, known as sclerotia, can develop on stems and it is these structures that survive over summer on lentil stubble. Severely infected leaves, flowers and pods wilt and fall to the ground. Plants ripen prematurely due to infection of the lower stem. If conditions remain wet throughout spring, stems can become girdled and patches of brown dying plants appear in the crop.

Infected pods will fail to fill properly and will produce shrivelled seed or no seed at all. Infected seed is discoloured.

Greyish/brown spots on leaflets of a botrytis grey mould infected lentil plant.

Disease Cycle

The disease inoculum readily survives on alternate hosts such as faba bean (Botrytis fabae) and chickpea (Botrytis cinerea), which are also grown in close rotation with lentil. The pathogens can survive on stubble residue (lentil, chickpea or faba bean) or in the soil as sclerotia but can also be carried on infected seed. When seed infected by these fungi are sown seedling blight will occur soon after crop emergence, reducing plant populations. Masses of spores are produced on infected plants which can then spread onto surrounding plants by wind and rainsplash to begin new infections.

Symptoms of grey mould appear more commonly in early spring. Dense crop canopies combined with warm, wet conditions in late winter and early spring favour development and spread of the disease. Moisture is essential for infection. Lentil crops sown early and/or at high seeding rates appear to be the worst affected by BGM. The disease usually starts off affecting the base of the stem. Prolonged wet conditions can cause it to progress higher up the plant subsequently affecting pods.

Management

Use Clean Seed

Only use seed with less than 5 percent botrytis infection; ideally use seed with nil infection. Using old or damaged seed can reduce seedling vigour and increase susceptibility to infection.

Varietal Resistance

Use varieties with greater resistance to botrytis grey mould. Warning – all varieties are susceptible to grey mould in areas and years where disease pressure is high.

Paddock Selection

Avoid planting this season's crop near old lentil, faba bean, chickpea, vetch or lathyrus stubble. These crop residues can harbour Botrytis fabae or Botrytis cinerea. A program of stubble reduction may also be undertaken by grazing or burying to reduce the carry- over of infected stubble into the following season. Allow a break of at least 3 years between lentil crops.

Seed Treatment

Use a registered seed treatment for the control of seed-borne diseases in lentil. Seed treatments can have a deleterious effect on rhizobia. Therefore, seed should be treated with fungicide and then inoculated with rhizobia in two separate operations. Rhizobia should be applied to seed immediately before sowing, especially on acid soils.

Sowing

Sow later and reduce seeding rates. Early sowing and high sowing rates can cause rank crop growth, lodging and increased risk of grey mould. Follow the recommended sowing rates and sowing dates for your district.

Use Foliar Fungicides

In areas of high risk it may be necessary to apply foliar fungicides to protect the crop, especially if a susceptible variety is being grown. Fungicides should be applied before canopy closure for best results. If conducive (warm and wet) conditions continue, follow up sprays may be necessary 12- 14 days later.

There are a range of fungicides available to control botrytis and selection of the most appropriate fungicide could depend on the level of disease pressure present. Fungicides containing mancozeb, chlorothalonil, carbendazim, or procymidone have activity against botrytis. If disease pressure is high then carbendazim or procymidone are the preferred fungicides. It is worth noting that these products are protectants and are most effective if applied before disease development.

COMMON SCAB

Common scab is a plant disease of root and tuber crops caused by a small number of Streptomyces species, specifically S. scabies, S. acidiscabies, S. turgidiscabies and others. Common scab mainly affects potato (Solanum tuberosum), but can also cause disease on radish (Raphanus sativus), parsnip (Pastinaca sativa), beet (Beta vulgaris), and carrot (Daucus carota). This plant disease is found wherever these vegetables are grown.

Common scab symptoms are variable and can range from surface russeting to deep pits in root and tuber vegetables. This disease does not usually affect yields, but it can greatly reduce quality of the harvested vegetables and make them unsuitable for sale.

Root and tuber vegetables are susceptible to infection by Streptomyces species as soon as the root or tuber forms, but, because this disease only affects root and tubers, the symptoms are not usually noted until harvest. Dry soils increase disease incidence and severity, therefore proper irrigation can aid in control of this disease. Common scab is suppressed if the soil pH is lower than 5.2, although scab lesions may still form in low pH soils due to physiological stresses or S. acidiscabies.

ASCOCHYTA BLIGHT OF CHICKPEAS

Ascochyta blight of chickpeas is caused by the fungal pathogen Phoma rabiei (formerly known as Ascochyta rabiei) which is specific to chickpeas. In recent years this disease has become the most important disease in chickpeas. In 1998 there was a serious outbreak of the disease in Victoria, South Australia and New South Wales which destroyed many crops. The extent of the chickpea cropping area fell considerably as a consequence of the varieties available being susceptible to ascochyta blight. Varieties now generally have much better foliar resistance to ascochyta blight. However, all varieties still need fungicides applied at flowering and podding stages as they are vulnerable to ascochyta blight during this growth stage.

Usually this disease becomes obvious in late winter when small patches of blighted plants appear throughout the paddock. The disease spreads during cool, wet weather in spring from infected plants to surrounding plants by rain splash of spores. This forms large blighted patches within crops.

Ascochyta blighted patches occur within a crop as the disease spreads from infected plants to surrounding healthy plants.

Typical leaf lesions caused by Ascochyta blight. These are seen as round spots with brown margins. Small dark specks (fruiting bodies) are often present.

Typical stem lesions caused by Ascochyta infection. Stems break at lesions and wither.

Ascochyta infected pod with a large lesion containing fruiting bodies.

Initially ascochyta blight appears on the younger leaves as small water-soaked pale spots. These spots rapidly enlarge, under cool and wet conditions, joining with other spots on the leaves and blighting the leaves and buds. Small black spots (pycnidia) can be seen in the affected areas. In severe cases the entire plant dries up quickly. Elongated lesions form on the stem and may girdle the stem which may then die and break off. Regrowth may occur from this point. Affected areas on the pods tend to be round, sunken, with pale centres and dark margins. The fungus can penetrate the pod and infect the seed.

When infected seeds are sown, the emerging seedlings will develop dark brown lesions at the base of the stem. Affected seedlings may collapse and die (damping-off).

Disease Cycle

This fungal disease has an asexual (non-sexual) and sexual stage; the asexual stage is most common in Australia. In this stage the fungus survives mainly on infected seed and on crop residues. Spores of the fungus produced on crop residues are carried into new crops by wind. Infection can occur at any stage of plant growth provided conditions are favourable. Moisture is essential for infection to occur. During wet weather, the disease can spread further than in dry conditions because spores of the fungus are carried onto neighbouring plants by wind and rain splash.

Management

Management of ascochyta blight requires a combination of farm hygiene, resistant varieties, crop monitoring and the use of fungicides. When growing a new variety, obtain a copy of the variety management package for information on specific disease management. A key strategy is to consider growing a resistant variety such as PBA Slasher or Genesi 090. Note - a fungicide application is still required at early podding to protect seed quality.

For chickpea varieties (rated susceptible to moderately resistant) the following management strategies should be applied:

- Use seed from a paddock where ascochyta was not detected or was well managed.

- Treat all seed with a seed dressing registered for ascochyta blight control.
- Choose a paddock at least 500 metres from last years chickpea crop.
- Generally the first fungicide spray will need to be applied 4 - 6 weeks after sowing. Moderately resistant varieties will require 2 - 4 strategic sprays throughout the growing season. The moderately susceptible varieties will require spraying every 2 - 3 weeks. Susceptible varieties will require spraying at least fortnightly throughout the growing season.
- Sprays only protect the plant parts contacted by the spray, subsequent plant growth will not be protected.
- A fungicide spray at pod set is essential to ensure production of premium quality seed.
- Always disinfect machinery, vehicles and boots once they have been in an infected crop.

Breeding programs in Australia are developing chickpea varieties with improved ascochyta blight resistance. All current resistant varieties and future releases will still require spraying at podding time, as their pods are susceptible to infection. However, the greatly reduced number of spray applications required for resistant varieties make chickpea a valid crop option and profitable to grow.

As resistant varieties are grown over larger areas, there is likely to be a reduction in the amount of ascochyta blight inoculum present. This should reduce the risk of further serious disease outbreaks.

BACTERIAL BLIGHT OF FIELD PEAS

This disease, caused by the bacteria Pseudomonas syringae pv. pisi and P. syringae pv. syringae, is a serious disease of peas. Recently, the pathogen P. syringae pv. syringae has been considered as the main cause of the disease in pea crops. Bacterial blight is widespread in field peas in southern New South Wales and Victoria, but its severity varies greatly from crop to crop and between seasons. The disease is seed borne and is more prevalent after frost events; multiple frosts can cause epidemics resulting in significant yield loss.

- Bacterial blight first appears as small, dark green, water-soaked spots on leaves and stipules, often near the leaf base. The spots enlarge and merge but are often limited by the veins.
- The leaf spots turn yellowish and later brown and papery.

- Spots on pods are sunken and olive brown.
- Spots can develop on the stem near ground level. These begin as water-soaked areas that later turn olive-green to dark brown. Stem lesions may coalesce, causing the stem to shrivel and die.
- Stem lesions may spread upwards to the stipules and leaflets. In this case, a fan-like lesion is formed on the stipule. Spots can merge causing the stem to shrivel and die.

Pre-emergence and post emergence damping-off may occur and even advanced plants may be killed. Heavily infected seed may be discoloured, but light infection has no visible effect on seed.

Leaf lesions caused by bacterial blight.

The symptoms of bacterial blight caused by Pseudomonas syringae pv. pisi or Pseudomonas syringae pv. syringae are indistinguishable from each other.

Water soaked lesion spreading into the leaf from the base caused by bacterial blight.

Disease Cycle

Bacterial blight commonly becomes established within a field by sowing infected seed or from infected pea trash that is nearby. During wet weather, bacteria spread from infected to healthy plants by rain splash and in wind-borne water droplets.

Infection may occur at any stage of plant growth. The pathogens can remain on the surface of plants without causing symptoms. However, following rain, heavy dew, frost or other forms of damage to plant tissues, symptoms can develop. Damage to field peas provides entry of bacteria into the plant tissue. Early infections may lead to epidemics but later infection can also cause yield losses. Because the disease depends on wet conditions, bacterial blight is most severe in wet seasons. A combination of excessive rainfall and strong winds provides the most favourable conditions for spread of the disease within crops.

Pseudomonas syringae pv. pisi is largely restricted to field peas but P. syringae pv. syringae has a wide host range including clover, common beans, faba beans, lentils, chickpeas and vetch which act as alternate hosts.

Management

Bacterial blight can be avoided by using an integrated approach to management that encompasses planting disease-free seed, crop rotation, variety selection and avoiding early sowing.

Use of Disease Free Seed

This is the main control measure recommended. The use of clean seed will minimise the possibility of disease, provided the land has not been cropped to peas for several years. Do not use seed from crops identified with bacterial blight during field inspections. A field inspection should occur at mid to late pod fill. Bacteria remain viable on seed for at least 2 years.

Rotations

To obtain a blight-free crop, peas should not be sown on land sown to peas in the previous year or adjacent to pea stubble. Where possible, peas should not be grown on the same land more than once in three years. If disease occurs the rotation should be extended to once in four years.

Stubble can be a significant source of inoculum. Destroy by burying, baling or burning infected stubble. The survival time of inoculum is significantly reduced by burying pea trash 10 cm below the soil surface.

Time of Sowing

Early sown crops are more vulnerable to bacterial blight infection than late sown crops; never sow earlier than recommended for your district. In areas prone to bacterial blight avoid early sowing.

Crop Damage

Bacterial blight is often associated with physical crop damage such as hail, frost, strong winds, sand blasting or machinery damage. Physical damage enables bacteria to enter plant tissue. Minimise the use of post emergence herbicide sprays, if possible, as the severity of bacterial blight can increase if plant tissue is damaged. Avoid paddocks where sulfonylurea residues may be present and the more frost-prone paddocks.

Varieties

The frequency of bacterial blight can be reduced by avoiding varieties susceptible to Pseudomonas syringae pv. syringae, and sowing those that have better resistance.

Farm Hygiene

When bacterial blight is detected, steps should be taken to prevent the spread of disease. Where possible, harvest infected crops last to avoid contaminating healthy crops and machinery used in an infected crop should be cleaned thoroughly and washed with disinfectant after use. Likewise, machine operators and farm workers should only move from crop to crop after taking precautions against the spread of bacteria. This is best achieved by wearing rubber boots and waterproof trousers that are washed with disinfectant immediately after leaving an infected field. Crops should never be inspected when they are wet as this increases the chance of spreading disease.

Chemical Control

Fungicides and seed treatments are designed to be active against fungal diseases and are ineffective in the control of bacterial diseases. There are copper based compounds, registered for use in field peas against bacterial blight, but evidence for their effectiveness in Australian field pea crops is limited and inconclusive.

SOIL BASED DISEASES

Rhizoctonia

Rhizoctonia is a genus of anamorphic fungi in the order Cantharellales. Species do not produce spores, but are composed of hyphae and sclerotia (hyphal propagules) and are asexual states of fungi in the genus *Thanatephorus*. *Rhizoctonia* species are saprotrophic, but are also facultative plant pathogens, causing commercially important crop diseases. They are also endomycorrhizal associates of orchids. The genus name was formerly used to accommodate many superficially similar, but unrelated fungi.

Taxonomy

Current Status

In its current sense, the genus is effectively restricted to the type species *R. solani* and its synonyms. The genus name is still, however, widely used in its old, artificial sense. Molecular research, based on cladistic analysis of DNA sequences, places *Rhizoctonia* within the family Ceratobasidiaceae.

Redisposition of Former Species

A comprehensive survey and redisposition of species names in *Rhizoctonia* was published in 1994 by Andersen & Stalpers. Only a few frequently used names are listed below. Many older names are of uncertain application or were never validly published, or both.

- Rhizoctonia bataticola = Macrophomina phaseolina (Botryosphaeriaceae),
- Rhizoctonia carotae = Fibulorhizoctonia carotae (teleomorph Athelia arachnoidea) (Atheliaceae),
- Rhizoctonia cerealis = not a validly published name (teleomorph Ceratobasidium cereale) (Ceratobasidiaceae),
- Rhizoctonia crocorum = Thanatophytum crocorum (teleomorph Helicobasidium purpureum) (Helicobasidiaceae),
- Rhizoctonia fragariae = not a validly published name (teleomorph Ceratobasidium cornigerum) (Ceratobasidiaceae),
- Rhizoctonia goodyerae-repentis = a name of uncertain application, but generally interpreted as Ceratobasidium cornigerum (Ceratobasidiaceae),
- Rhizoctonia leguminicola = Botrytis fabae (Sclerotiniaceae),
- Rhizoctonia oryzae = not a validly published name and of uncertain application, but generally interpreted as Waitea circinata (Corticiaceae),
- Rhizoctonia ramicola = Ceratorhiza ramicola (teleomorph Ceratobasidium ramicola) (Ceratobasidiaceae),
- Rhizoctonia rubi = not a validly published name and of uncertain application (probably Ascomycota),
- Rhizoctonia zeae = Waitea circinata (Corticiaceae).

Habitat and Distribution

Species are saprotrophic, occurring in the soil and producing basidiocarps (fruit bodies of the Thanatephorus teleomorph) on dead stems and plant detritus. They are also opportunistic plant pathogens, with an almost unlimited host range, and have been isolated from orchid mycorrhiza. Distribution appears to be cosmopolitan.

Economic Importance

Rhizoctonia solani causes a wide range of commercially significant plant diseases. It is one of the fungi responsible for Brown patch (a turfgrass disease), damping off in seedlings, as well as black scurf of potatoes, bare patch of cereals, root rot of sugar beet, belly rot of cucumber, sheath blight of rice, and many other pathogenic conditions.

An efficient conversion of tryptophan to indole-3-acetic acid (IAA) and tryptophol can be achieved by some species in the genus *Rhizoctonia.*

Verticillium

Verticillium dahliae infecting sunflower.

Verticillium is a genus of fungi in the division Ascomycota, and are an anamorphic form of the family Plectosphaerellaceae. The genus used to include diverse groups comprising saprobes and parasites of higher plants, insects, nematodes, mollusc eggs, and other fungi, thus the genus used to have a wide-ranging group of taxa characterised by simple but ill-defined characters. The genus, currently thought to contain 51 species, may be broadly divided into three ecologically based groups - mycopathogens, entomopathogens, and plant pathogens and related saprotrophs. However, the genus has undergone recent revision into which most entomopathogenic and mycopathogenic isolates fall into a new group called Lecanicillium. The genus now includes the plant-pathogenic species V. dahliae, V. longisporum, V. albo-atrum, V. nubilum, and V. tricorpus.

The better-known species of Verticillium are V. dahliae, V. albo-atrum and V. longisporum, which cause wilt diseases called verticillium wilts in more than 400 eudicot plant species.

Selected Species

- Verticillium dahliae Kleb.

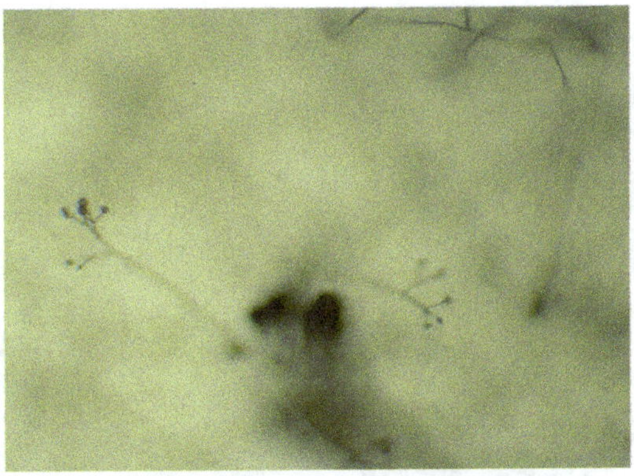

Verticillium conidiophores.

- Verticillium albo-atrum Reinke & Berthold - causes verticillium wilt or maple wilt. First identified from potatoes in Germany in 1870, this species attacks over 300 different cultivated plants and can persist as a saprotrophic soil organism for more than 15 years. When infecting ornamental trees such as maples, elms, aspen, ash, beech, catalpa, oak, and others, the first symptoms are midsummer wilting on one side of a tree or branch. The sapwood has greenish or brownish streaks, and the infection can take a few years to progress to the rest of the tree or move rapidly. The fungi universally move up the xylem vessels. In fruit trees, the infection is known as black heart, and is common in apricots and sometimes affects almond, peach, plum, and avocado trees. This fungus also affects herbaceous ornamentals and vegetables such as chrysanthemums, mints, *Lychnis* spp., tomatoes, eggplants, okra, and rhubarb, causing wilting and death. Identification can be made by looking for one-celled conidia, hyaline round to ellipsoid which are formed at the tips of whorled branches. They are easily separated from the tips.
- Nematode pathogens, previously in *Verticillium*, have now been placed in the new genus *Pochonia*.
- All insect pathogens have been placed in the new genus *Lecanicillium*. An approved name of *Verticillium lecanii* (Zimmerman) Viegas is now *Lecanicillium lecanii* although isolates may belong to: *L. attenuatum, L. longisporum, L. muscarium,* or *L. nodulosum*.

Crown Rot

The symptoms of crown rot (caused by Fusarium pseudograminearum or Fusarium culmorum) are most obvious on plants close to maturity. However, the fungus may cause a seedling blight which can result in either pre or post emergent death. Plants affected by crown rot are frequently stunted and produce fewer tillers. Disease symptoms at the base of plants include honey brown discolouration of the crown, lower leaf sheaths and tillers. Symptoms can extend up the stem where the fungus may form pink spore masses at the nodes. By comparison, plant roots and crowns infected with take-all are distinctly black in colour.

Honey discolouration on stem caused by the crown rot fungus.

Honey discolouration stem and pink fungal growth caused by the crown rot fungus.

When the disease is severe, plants die prematurely and deadheads or whiteheads are produced. The deadheads are either empty or partially filled with shrivelled grain. Affected plants may be scattered across a paddock or occur in patches. Yield losses are greatest in dry years, especially when moisture stress occurs after flowering.

Crown rot causes deadhead that are usually found on scattered plants or single tillers in the crop.

Disease Cycle

The fungus can survive for up to two years on infected cereal stubble from previous cereal plants, volunteer plants or grass weeds. For this reason, stubble management is an important aspect in the control of crown rot. Survival is enhanced by dry summer conditions, but disease development is favoured by moist humid conditions with temperatures between 15 °C and 20 °C.

Management

Reducing crown rot inoculum is the key to controlling this disease and rotations are the most important component in the integrated management of crown rot.

Rotations with non-susceptible crops will reduce the severity of crown rot. A two year break with a pulse, oilseed crop or fallow will reduce crown rot following wheat or barley crops. Grass-free lucerne and medics will also reduce disease severity. Some varieties have partial resistance to crown rot and can reduce the severity of crown rot. However, most bread wheat, durum wheat and barley varieties are either susceptible or highly susceptible to crown rot.

Stubble management is an important factor determining the survival of the fungus. Agronomic practices that encourage greater retention or slower decomposition of stubble can increase the severity of crown rot. Tillage helps break down infected stubble by accelerating its decomposition thereby reducing inoculum levels.

Burning stubble can reduce, but not control crown rot. However, burning stubble is undesirable as the retention of stubble helps reduce erosion and aids the storage of water over summer. As a compromise solution, burn stubble as close as possible to sowing.

Infection rates can be reduced by sowing between rows of standing stubble. In the southern region inter- row sowing using accurate GPS autosteer has shown a decrease in the number of infected plants by 50 percent – resulting in a 5 to 10 percent yield increase.

There are no fungicides currently available to control crown rot.

Hosts

Crown rot can be a serious problem in durum wheat, as well as bread wheat, barley and oats. Other host plants include wild oats, canary grass (phalaris), wheat grass, brome grass, barley grass, winter grass and ryegrass.

References

- Crop-diseases-the-nightmare-of-every-farmer, post: agrivi.com, Retrieved 14 July, 2019
- "Botrytis Fruit Rot / Gray Mold on Strawberry | NC State Extension Publications". Content.ces.ncsu.edu. Retrieved 2017-12-11
- Crop-disease-manual: croppro.com.au, Retrieved 17 May, 2019
- "UC IPM: UC Management Guidelines for Botrytis Diseases And Disorders on Citrus". Ipm.ucanr.edu. Retrieved 2017-12-11
- Wheat-rust-diseases phcontent-6-divaccordion, learning-center: cropscience.bayer.us, Retrieved 19 April, 2019
- Lerat, S.; Simao-Beaunoir, A. M.; Beaulieu, C. (2009). "Genetic and physiological determinants of Streptomyces scabies pathogenicity". Molecular Plant Pathology. 10 (5): 579–85. Doi:10.1111/j.1364-3703.2009.00561.x. PMID 19694949
- Rhizoctonia diseases of sugar beet "Archived copy". Archived from the original on 2010-06-19. Retrieved 2010-08-05
- Kirk PM, Cannon PF, Minter DW, Stalpers JA (2008). Dictionary of the Fungi (10th ed.). Wallingford: CABI. P. 724. ISBN 978-0-85199-826-8

Crop Protection

CHAPTER 5

The science and practice of managing plant diseases and pests that harm agricultural crops is referred to as crop protection. Some of the chemicals and biological compounds that are used to manage crop are fungicides, insecticides, biopesticides, etc. The chapter closely examines these chemicals used in crop protection to provide an extensive understanding of the subject.

Crop protection is the practice of removing weeds, adjusting weather and eliminating pests. It may include managing everything that inhibit the growth of fruit, vegetables and all farm produce.

It is highly important because through this, higher quality crops are produced and wastage is reduced. The increased crop production through proper crop protection leads to less use of resources like land, water and labor. With less usage of land, biodiversity is preserved. Through this, more farm crops and produce actually reach the markets, keeping the prices low.

If farmers were to stop methods of crop protection, according to UN Food and Agricultural Organization (FAO), around 40 percent will be lost due to pests and plant diseases.

An approach called integrated Pest Management (IPM) is helpful in reducing the wastage of farm resources. In IPM, they combine the best cultural, biological and chemical measures to keep pest and diseases away from the crops.

There are several ways and approaches to successful plant protection:

Chemical Intervention

First, we use herbicides to kill unwanted plants and weeds. Weeds can be a real problem because they contest with the food crops for nutrients and space.

Plant-eating insects and roundworms attack the farm crops leaving these damaged and inedible. To eliminate these harmful insects, we use insecticides.

Lastly, we use fungicides for fungi that cause serious disease and damage to agriculture.

Biological Pest Control

With biological means of pest control, we manage weeds, insects and diseases by using other living organisms. We make use of beneficial insects that perform these valued services. You may refer to this link to learn more about beneficial insects.

Barrier Methods

We can also opt for this method in which we create barriers between plants and potential harmful pests, weather and diseases. We have bird netting, micromesh and the likes. We can also employ electronic pest deterrents.

HERBICIDES

Herbicide is an agent, usually chemical, for killing or inhibiting the growth of unwanted plants, such as residential or agricultural weeds and invasive species. A great advantage of chemical herbicides over mechanical weed control is the ease of application, which often saves on the cost of labour. Most herbicides are considered nontoxic to animals and humans, but they can cause substantial mortality of nontarget plants and the insects that depend on them, especially when applied aerially.

Chemical weed control has been used for a very long time: sea salt, industrial by-products, and oils were first employed. Selective control of broad-leaved weeds in fields of cereal crops was discovered in France in the late 1800s, and this practice soon spread throughout Europe. Sulfates and nitrates of copper and iron were used, and sulfuric acid proved even more effective. Application was by spraying. Soon sodium arsenite became popular both as a spray and as a soil sterilant. On thousands of kilometres of railroad right-of-way, and in sugarcane and rubber plantations in the tropics, the hazardous material was used in tremendous quantities, often resulting in the poisoning of animals and occasionally humans.

Sinox, the first major organic chemical herbicide, was developed in France in 1896. In the late 1940s new herbicides were developed out of the research during World War II, and the era of the "miracle" weed killers began. Within 20 years over 100 new chemicals were synthesized, developed, and put into use. Chemical weed control superseded both plant-disease and insect-pest control in economic impact. In particular, the year 1945 was key to the development of selective chemical weed control. Introduced then were 2,4-D (2,4-dichlorophenoxyacetic acid), 2,4,5-T (2,4,5-trichlorophenoxyacetic acid), and IPC (isopropyl-N-phenylcarbamate)—the first two selective as foliar sprays against broad-leaved weeds, the third selective against grass species when applied through the soil.

The new herbicides were revolutionary in that their high toxicity allowed for effective weed control at dosage rates as low as one to two kilograms per hectare (one or two pounds per acre). This contrasted with carbon bisulfide, borax, and arsenic trioxide, which were required at rates of up to 2,242 kilograms per hectare (one ton per acre), and with sodium chlorate, required at rates of around 112 kilograms per hectare (100 pounds per acre). However, some of those early herbicides, including 2,4,5-T, were later deemed unsafe for humans and the environment and were discontinued in many countries. Effective herbicides have continued to be developed, and some, such as glyphosate, are widely used around the world.

Since the mid-1980s, certain agricultural plants, known as herbicide-resistant crops (HRCs), have been genetically engineered for resistance to specific chemical herbicides, notably glyphosate. These genetically modified organisms (GMOs) enable effective chemical control of weeds, since only the HRC plants can survive in fields treated with the corresponding herbicide. Such crops have been especially valuable for no-till farming, which helps prevent soil erosion. However, because these crops encourage increased application of chemicals to the soil rather than decreased application, they remain controversial with regard to their environmental impact and general safety. In addition, in order to reduce the risk of selecting for herbicide-resistant weeds, farmers must use multiple diverse weed-management strategies.

Genetically modified corn (maize).

Application

Modern weed killers are put in two categories: selective (affecting specific plant species) and nonselective (affecting plants generally). These, in turn, are classified as foliage-applied and soil herbicides. Contact herbicides (e.g., sulfuric acid, diquat, paraquat) kill only the plant organs with which they are in contact. Translocated herbicides (e.g., amitrole, picloram, and 2,4-D) are effective against roots or other organs to which they are transported from aboveground treated surfaces (i.e., soil). With respect to planting time, herbicides are also classified as preplant, preemergence, or postemergence weed killers. Preplant herbicides may be applied to the soil or to weeds before crop planting.

Many millions of hectares are treated from the air each year, particularly cereal croplands, pastures, rangelands, forests, and other situations where an airplane can be used. Modern equipment for treating row cropland with herbicides has made weed control increasingly convenient. Sprayers, soil incorporation equipment, and spreaders for pelleted herbicides have all added to the convenience of, and removed uncertainty from, herbicide application. Machinery is available that simultaneously builds up beds, plants the seed, sprays with insecticide, and incorporates fertilizer and pre-emergence herbicide all in one operation.

Pesticides, Farm tractor spraying pesticides on crops.

Certain herbicides (e.g., sodium arsenite) are sometimes applied by the jar method, whereby the tops of weeds are bent over and immersed in jars of poisonous solution. The herbicide is drawn into the rest of the plant and into connecting plants, gradually killing the entire system. Wild morning glory, poison oak, and camel thorn are sometimes treated in this manner. Chlorinated benzene and certain aromatic solvents are used to control aquatic weeds by adding them directly to the watery medium.

FUNGICIDES

A fungicide is any compound that is toxic to fungi. A fungicide can work to control disease in one or a combination of ways: from inhibiting spore germination, hyphal growth, or limiting development of spores. Fungicides may be fungistatic, meaning they prevent new growth or sporulation, or fungicidal, meaning they kill the fungus outright.

Fungicides can be applied to the seed, fertiliser, soil or foliage. The most common uses for field crops are as seed or foliar treatments.

Any fungicide used in Australia must be registered with the Australian Pesticides and Veterinary Medicines Authority (APVMA). By using only registered agricultural chemicals the user, industry and general public can be assured that, if the chemical is used in accordance with the label, the risks posed by that use in terms of public health, trade and the environment are minimised.

Seed, Fertilizer and Foliar Applications of Fungicides

Seed Treatments

Fungicide seed treatments protect seed viability and inhibit diseases like seed rot and seedling blight. Seed treatments protect the seed either by controlling fungi present on the seed surface or carried internally in the seed, or by controlling fungi present in the soil or on crop residue in the soil. Some treatments will protect a young seedling against early leaf disease or root rot infection, but in most cases, seed treatments are no longer effective after seedling emergence. A seed treatment can also reduce the potential for introducing a pathogen into an area where it is not established:

- In general, seed treatments may have either systemic or contact activity. A systemic product is required to control fungi carried within the d's embryo, cotyledons or seed coat, (i.e. smut in barley, ascochyta in lentil). A systemic seed product is also needed to control early seedling diseases such as scald in barley, stripe rust in wheat, or early season infection of ascochyta in lentil. A contact or protectant product is adequate for surface-borne or soil-borne fungi.

- The use of seed treatments is becoming a common practice because treating seed ensures the crop gets off to a good start and is relatively 'cheap' insurance. Keep in mind that seed treatments will not 'cure' a poor seed lot that has high proportions of dead, damaged or infested seed.

Fertilizer Treatments

The application of fungicides has become increasingly used in cereal crops during recent years. Fungicides applied in these can reduce the severity of the root disease take-all and give long term control of stripe rust. It is important to remember that fertiliser-applied fungicides do not control bunts and smuts so seed applied fungicides are still required.

Foliar Fungicides

Foliar fungicides are an important tool to protect the above ground plant from infection by disease pathogens. The goal of disease management is not necessarily to eradicate the disease, which is impossible, but to reduce it to levels below the economic threshold:

- Routine applications of fungicide are generally not economical. Both the fungicide product and its application are a significant cost to the farmer, so it is important to understand the disease cycles. Use proper scouting techniques to determine disease risk.

- Fungicides are most effective if applied in a preventative manner, either at the first sign of disease (for polycyclic diseases) or if disease development is forecasted (for monocyclic diseases).

- Unfortunately, there are limited decision support systems or forecasting tools available to determine when the economic threshold for diseases is reached and when a fungicide application is warranted.

Residues

In recent years an increase in the rate of detection of unacceptable fungicide residues has been noted in industry funded residue monitoring programs, eg. National Residue Survey. The increase is strongly associated with lack of decontamination of equipment used to store, transport and handle fungicide treated fertiliser, particularly in relation to flutriafol.

Traceback investigations conducted by DEDJTR have shown that use of silos, augers and bins to store flutriafol treated fertiliser is directly linked to unacceptable residues. As there are no instructions on how to adequately decontaminate such equipment once it has been used to handle flutriafol treated fertiliser, it is critical that growers have separate equipment for grain handling. Whilst this may seem expensive it will avoid the grower selling contaminated produce, which can jeopardise market access for all Australian growers if detected by an export market. It is also an offence under the Agricultural and Veterinary Chemicals (Control of Use) Act 1992 for a grower to sell contaminated produce.

Traceback investigations have also shown a high likelihood of some residues resulting from backloading of grain in trucks used by contractors to transport flutriafol treated fertiliser. Whilst current industry codes of practices may attempt to instruct drivers on how to decontaminate their trucks, there are no instructions on the label of flutriafol products.

Growers are advised to avoid backloading grain into trucks that have been used to transport flutriafol treated fertiliser, regardless of the drivers best efforts and assurances. In this situation the grower may still be held responsible for selling contaminated produce.

Mode of Action

The manner in which the fungicide affects the fungus is called the mode of action. Fungicides are divided into chemical groups based on their mode of action.

Most of the fungicides used in pulse and cereal crops in Victoria fall into five main chemical groups based on their modes of action, as follows:

Benzimidazoles

These products provide protective and curative control and show apoplastic mobility. Mode of action is by inhibiting mitosis, more specifically, by binding the tubulin needed for cell and nuclear division thus resulting in death of the fungus. Examples include the active ingredients carbendazim and thiabendazole, which are used as pulse fungicides.

Carbendazim is an S7 Dangerous Poison and can only be used on pulse crops in strict accordance with the product label.

Dicarboximides

Many of the pulse fungicides fall within this group, active ingredients iprodione and procymidone. Iprodione is a contact fungicide with both preventive and curative activity. Procymidone is an S7 Dangerous Poison and can only be used on cereal crops in strict accordance with the product label. It is a systemic fungicide with both preventative and eradicant activity. These two chemicals act on the fungi by inhibiting spore germination and affecting cell division. The benefits of these products are that they control a unique group of fungi including botrytis and sclerotinia.

DMI - Triazoles

All of the fungicides registered for cereal crops are in this group; with the exception of products which also contain a strobilurin (Group 11) (i.e. cyproconazole and azoxystrobin). Triazoles are de-methylation inhibitors (DMI) which prevent ergosterol biosynthesis. Triazoles have systemic activity and provide protective, curative and eradicant control. Examples include the active ingredients propiconazole, tebuconazole, triadimenol, triadimefon, triticonazole, and epoxyconazole.

SDHI (Succinate Dehydrogenase Inhibitors)

SDHIs are locally systemic, movement is translaminar and upward. In Australia, SDHIs are mainly used as seed treatments either on their own or in combinations with other chemical groups. Examples include carboxin (Group 7), penflufen (Group 7), fluxapyroxad (Group 7) or seed treatments combining difenoconazole (Group 3), metalaxyl-m (Group 4) and sedaxane (Group 7). Foliar fungicide formulations have been released overseas, but resistance has built up quickly.QOLs contain the strobilurins which are preventative and locally systemic (translaminar).

QoI (Quinone Outside Inhibitor)

Some QoIs, like azoxystrobin can move in the xylem. They are generally mixed with triazoles for agricultural use in Australia to reduce the chance of resistance.

Licencing and use of Chemicals

Under the Agricultural and Veterinary Chemicals (Control of Use) Act 1992 any person that uses restricted use chemicals such as S7 Dangerous Poisons must hold appropriate authorisation, such as an Agricultural Chemical Users Permit issued by DEDJTR. As noted, this applies to some commonly used fungicides such as carbendazim and procymidone.

All restricted use chemicals must be used in strict accordance with the label, and the user must keep the required records of use of the chemicals.

It is illegal to use unregistered chemicals, and it is also illegal for a grower to sell agricultural produce that contains unacceptable chemical residues.

Off label use of restricted use chemicals in Victoria is not permitted for restricted use chemicals unless a permit has been obtained from DEDJTR.

Off label use of other chemicals may be permitted as long the chemical is not used:

- At a rate greater than the maximum label rate,
- At a frequency greater than the label frequency,
- Contrary to any relevant prohibitive label statements ('DO NOT statements').

A permit is required from the Australian Pesticide and Veterinary Medicines Authority (APVMA) to contravene any of these restrictions.

Any person considering off label use must address any risks posed in terms of residues (particularly in terms of export market requirements), occupational health and safety and environment.

Fungicide Resistance

- The fungicides within each chemical group affect the same target sites. Over time, strains of the fungus can develop which are immune to the effects of the fungicide. This is known as fungicide resistance. Hence, it is important to rotate fungicides from different chemical groups (mode of actions) so that resistance will not be selected.
- If there is resistance to one fungicide within a chemical group it is likely that other fungicides within that group could also be present. This is known as cross resistance.
- The loss of a fungicide due to resistance would be costly to farmers because of increased disease levels and the loss of a product when no alternatives may be present. Therefore, we are all responsible to ensure judicious use of fungicides and to rotate between chemical groups or use a fungicide that contains active ingredients from more than one mode of action group.

Types of Fungicide Activity (Against the Fungal Target)

Fungicides may have one or more of the following types of activity:

Protective/Preventative Activity

Inhibits spore germination or kills germinating spores on the plant surface. Protectant fungicides must be applied to the plant prior to the arrival of the pathogen, or at least before it has a chance to germinate and enter its host. Coverage is very important because the fungicide must be in contact with the spore to be effective; hence new tissues from actively growing plants will not be protected. The majority of fungicides used in field crops in Victoria are in this category.

Curative Activity

Inhibits fungal growth prior to symptoms appearing but can be effective on fungi once they've infected the host. Curative activity is limited to the early part of the incubation period, likely only for the first 24 to 36 hours after spore germination and infection. Curative products do not repair tissue that has already been damaged or killed by the fungus. Many fungicides fall within this category.

Eradicant Activity

These fungicides have the ability to limit growth and spore production even after symptoms are visible. Few fungicides fall within this category. Examples include those active ingredients in the triazole family, e.g. propiconazole, tebuconazole, and triadimefon.

The goal for any fungicide is to protect plants from infection rather than cure them. Even if a fungicide does have some curative or eradicant properties, they are most effective if applied in a preventative manner.

Types of Movement within the Plant

In addition to the broad categories of activity listed above, fungicides can also be defined by their movement within the plant:

Contact Fungicides

Contact fungicides provide a barrier on the plant surface to prevent spores from germinating or entering. Minimal redistribution of the product occurs through the vapour phase or by rain, but there is no uptake of the chemical into the plant. Contact fungicides have protective / preventative activity only, therefore good coverage is essential. Examples include the pulse foliar fungicides mancozeb, chlorothalonil and the various copper-based products. Also, many of pulse seed treatments, e.g. the active ingredient thiram.

Systemic Fungicides

Systemic fungicides show some movement from the plant surface, into the plant and within the plant. Hence systemic fungicides can have preventative, curative and eradicant ability. Most are only locally systemic and not all are created equal:

- Some systemics have what is called apoplastic movement, meaning there is movement within the free space around cells, in cell walls and xylem (water-conducting) elements.

- Some systemics have acropetal movement, meaning movement towards the leaf tip and margins.
- No systemic fungicides are currently registered that have true systemic activity, also known as symplastic activity, which includes movement into the phloem, (i.e. As for systemic herbicides).

Refer to Figure, which shows a diagrammatic view of the infection process of ascochyta blight in chickpea. Contact fungicides are effective only at Stage #1; preventing the spore from germinating on the leaf surface, (i.e. chlorothalonil, mancozeb). Once inside the plant, the fungus is protected from dry conditions and the effects of most fungicides (Stage #2 and #3). Those fungicides with systemic or curative activity may have some activity against the fungus at this stage, as long as it is applied within 24 to 36 hours after infection (e.g. boscalid, azoystrobin, pyraclostrobin).

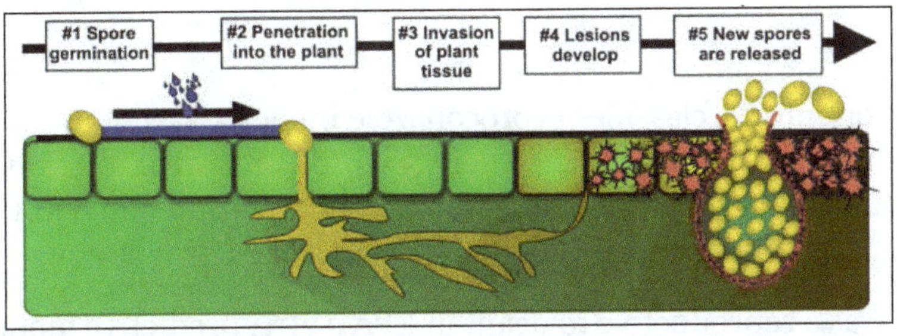

Diagram of the ascochyta blight infection process in chickpea (leaf in cross- section).

Cyproconazole

Cyproconazole is an agricultural fungicide of the class of azoles, used on cereal crops, coffee, sugar beet, fruit trees and grapes, on sod farms and golf courses and on wood as a preservative. It was introduced to the market by then Sandoz in 1994.

Mechanism of Action

Cyproconazole inhibits demethylation, a particular step in the synthesis of a component of the fungal cell wall called sterol. This means it affects fungal growth, but not the fungal sporulation. This explains why it must be used when fungal growth is maximum, early in the infection, because in late infections fungal growth slows down and the agent is ineffective.

Use

Formulations

Many different formulations exist with imazalil, difenoconazole, prochloraz, propiconazole, chlorothalonil, cyprodinil, fludioxonil, azoxystrobin, and copper. In wood preservatives it is mixed with didecyldimethylammonium chloride. It is the active ingredient in two foliar fungicides for soybeans in the U.S., Alto X, and mixed with azoxystrobin in Quadris Xtra, both by Syngenta. It is also manufactured by Bayer CropScience and Dow AgroSciences.

Application

Cyproconazole is used against powdery mildew, rust on cereals and apple scab, and applied by air or on the ground to cereal crops, coffee, sugar beet, fruit trees and grapes. It controls the following pests: Puccinia graminis, Puccinia spp., Pseudocercosporella herpotrichoides and Septoria species. It can be used on above-ground wood to prevent it from decay from fungi as an alternative to Chromated Copper Arsenate. It was originally marketed for use on sod farms and golf courses In the U.S., chemigation is allowed with less than half inch application, aerial spraying with a 5 gpa minimum, ground application is adequate for coverage and canopy penetration. The re-rentry interval is 12 hours. Reapplication within 30 days of harvest is not permitted.

Hazards

The European Community classifies cyproconazole into carcinogen category 3 as limited evidence, into the reproduction risk category 3 as "possible risk of harm to the unborn child", as harmful if swallowed, and dangerous for the environment, because very toxic to aquatic organisms, possibly causing long-term adverse effects in the aquatic environment.

Cyproconazole was a reproductive toxicant in EPA's 2011 predictive model of reproductive toxicity using Toxcast high-throughput screening. Cyproconazole like fluconazole, a triazole used in human medicine, can induce liver swelling in mice. As of 2015 the constitutive androstane receptor has been shown to be mediator for this effect. Cyproconazole has been shown to cause a dose dependent inhibition of progesterone production in human placental cells in vitro.

Toxicokinetics

Cyproconazole as other triazoles inhibits the enzyme cytochrome P-450, so it can no longer demethylate lanosterol, an intermediate needed in ergosterol synthesis. In fish, CYP mediated steroid metabolism and xenobiotic metabolism can be affected in opposite ways. The half-life of cyproconazole in trout was about 1.0 day.

Resistance

Development of fungal resistance can be prevented by not using cyproconazole "repeatedly alone in the same season" or by not using it late in the infection, that is, curatively. Fungi can develop resistance if the same fungicide is used repeatedly or when fungicides with the same mode of action are repeatedly.

INSECTICIDES

Insecticides are substances used to kill insects. They include ovicides and larvicides used against insect eggs and larvae, respectively. Insecticides are used in agriculture, medicine, industry and by consumers. Insecticides are claimed to be a major factor behind the increase in the 20th-century's agricultural productivity. Nearly all insecticides have the potential to significantly alter ecosystems; many are toxic to humans and/or animals; some become concentrated as they spread along the food chain.

Insecticides can be classified into two major groups: systemic insecticides, which have residual or long term activity; and contact insecticides, which have no residual activity.

Furthermore, one can distinguish three types of insecticide. 1. Natural insecticides, such as nicotine, pyrethrum and neem extracts, made by plants as defenses against insects. 2. Inorganic insecticides, which are metals. 3. Organic insecticides, which are organic chemical compounds, mostly working by contact.

The mode of action describes how the pesticide kills or inactivates a pest. It provides another way of classifying insecticides. Mode of action is important in understanding whether an insecticide will be toxic to unrelated species, such as fish, birds and mammals.

Insecticides may be repellent or non-repellent. Social insects such as ants cannot detect non-repellents and readily crawl through them. As they return to the nest they take insecticide with them and transfer it to their nestmates. Over time, this eliminates all of the ants including the queen. This is slower than some other methods, but usually completely eradicates the ant colony.

Insecticides are distinct from non-insecticidal repellents, which repel but do not kill.

Type of Activity

Systemic insecticides become incorporated and distributed systemically throughout the whole plant. When insects feed on the plant, they ingest the insecticide. Systemic insecticides produced by transgenic plants are called plant-incorporated protectants (PIPs). For instance, a gene that codes for a specific *Bacillus thuringiensis* biocidal protein was introduced into corn (maize) and other species. The plant manufactures the protein, which kills the insect when consumed.

Contact insecticides are toxic to insects upon direct contact. These can be inorganic insecticides, which are metals and include the commonly used sulfur, and the less commonly used arsenates, copper and fluorine compounds. Contact insecticides can also be organic insecticides, i.e. organic chemical compounds, synthetically produced, and comprising the largest numbers of pesticides used today. Or they can be natural compounds like pyrethrum, neem oil etc. Contact insecticides usually have no residual activity.

Efficacy can be related to the quality of pesticide application, with small droplets, such as aerosols often improving performance.

Biological Pesticides

Many organic compounds are produced by plants for the purpose of defending the host plant from predation. A trivial case is tree rosin, which is a natural insecticide. Specifically, the production of oleoresin by conifer species is a component of the defense response against insect attack and fungal pathogen infection. Many fragrances, e.g. oil of wintergreen, are in fact antifeedants.

Four extracts of plants are in commercial use: pyrethrum, rotenone, neem oil, and various essential oils.

Other Biological Approaches

Plant-Incorporated Protectants

Transgenic crops that act as insecticides began in 1996 with a genetically modified potato that produced the Cry protein, derived from the bacterium Bacillus thuringiensis, which is toxic to beetle larvae such as the Colorado potato beetle. The technique has been expanded to include the use of RNA interference RNAi that fatally silences crucial insect genes. RNAi likely evolved as a defense against viruses. Midgut cells in many larvae take up the molecules and help spread the signal. The technology can target only insects that have the silenced sequence, as was demonstrated when a particular RNAi affected only one of four fruit fly species. The technique is expected to replace many other insecticides, which are losing effectiveness due to the spread of pesticide resistance.

Enzymes

Many plants exude substances to repel insects. Premier examples are substances activated by the enzyme myrosinase. This enzyme converts glucosinolates to various compounds that are toxic to herbivorous insects. One product of this enzyme is allyl isothiocyanate, the pungent ingredient in horseradish sauces.

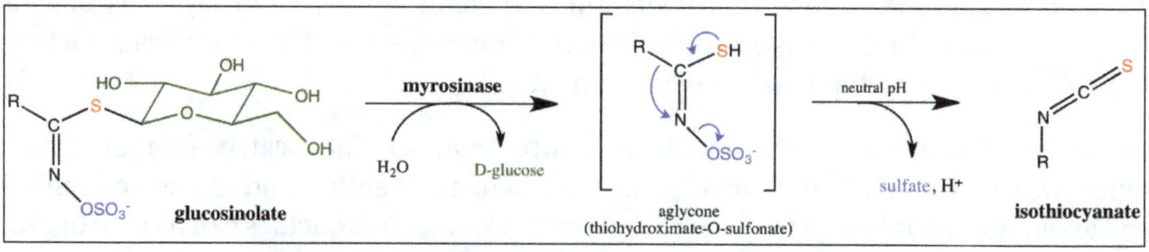

Biosynthesis of antifeedants by the action of myrosinase.

The myrosinase is released only upon crushing the flesh of horseradish. Since allyl isothiocyanate is harmful to the plant as well as the insect, it is stored in the harmless form of the glucosinolate, separate from the myrosinase enzyme.

Bacterial

Bacillus thuringiensis is a bacterial disease that affects Lepidopterans and some other insects. Toxins produced by strains of this bacterium are used as a larvicide against caterpillars, beetles, and mosquitoes. Toxins from Saccharopolyspora spinosa are isolated from fermentations and sold as Spinosad. Because these toxins have little effect on other organisms, they are considered more environmentally friendly than synthetic pesticides. The toxin from B. thuringiensis (Bt toxin) has been incorporated directly into plants through the use of genetic engineering. Other biological insecticides include products based on entomopathogenic fungi (e.g., Beauveria bassiana, Metarhizium anisopliae), nematodes (e.g., Steinernema feltiae) and viruses (e.g., Cydia pomonella granulovirus).

Synthetic Insecticide and Natural Insecticides

A major emphasis of organic chemistry is the development of chemical tools to enhance agricultural productivity. Insecticides represent a major area of emphasis. Many of the major insecticides are inspired by biological analogues. Many others are completely alien to nature.

Organochlorides

The best known organochloride, DDT, was created by Swiss scientist Paul Müller. For this discovery, he was awarded the 1948 Nobel Prize for Physiology or Medicine. DDT was introduced in 1944. It functions by opening sodium channels in the insect's nerve cells. The contemporaneous rise of the chemical industry facilitated large-scale production of DDT and related chlorinated hydrocarbons.

Organophosphates and Carbamates

Organophosphates are another large class of contact insecticides. These also target the insect's nervous system. Organophosphates interfere with the enzymes acetylcholinesterase and other cholinesterases, disrupting nerve impulses and killing or disabling the insect. Organophosphate insecticides and chemical warfare nerve agents (such as sarin, tabun, soman, and VX) work in the same way. Organophosphates have a cumulative toxic effect to wildlife, so multiple exposures to the chemicals amplifies the toxicity. In the US, organophosphate use declined with the rise of substitutes.

Carbamate insecticides have similar mechanisms to organophosphates, but have a much shorter duration of action and are somewhat less toxic.

Pyrethroids

Pyrethroid pesticides mimic the insecticidal activity of the natural compound pyrethrum, the biopesticide found in pyrethrins. These compounds are nonpersistent sodium channel modulators and are less toxic than organophosphates and carbamates. Compounds in this group are often applied against household pests.

Neonicotinoids

Neonicotinoids are synthetic analogues of the natural insecticide nicotine (with much lower acute mammalian toxicity and greater field persistence). These chemicals are acetylcholine receptor agonists. They are broad-spectrum systemic insecticides, with rapid action (minutes-hours). They are applied as sprays, drenches, seed and soil treatments. Treated insects exhibit leg tremors, rapid wing motion, stylet withdrawal (aphids), disoriented movement, paralysis and death. Imidacloprid may be the most common. It has recently come under scrutiny for allegedly pernicious effects on honeybees and its potential to increase the susceptibility of rice to planthopper attacks.

Ryanoids

Ryanoids are synthetic analogues with the same mode of action as ryanodine, a naturally occurring insecticide extracted from *Ryania speciosa* (Flacourtiaceae). They bind to calcium channels in cardiac and skeletal muscle, blocking nerve transmission. The first insecticide from this class to be registered was Rynaxypyr, generic name chlorantraniliprole.

Insect Growth Regulators

Insect growth regulator (IGR) is a term coined to include insect hormone mimics and an earlier class of chemicals, the benzoylphenyl ureas, which inhibit chitin (exoskeleton) biosynthesis in insects Diflubenzuron is a member of the latter class, used primarily to control caterpillars that are pests. The most successful insecticides in this class are the juvenoids (juvenile hormone analogues). Of these, methoprene is most widely used. It has no observable acute toxicity in rats and is approved by World Health Organization (WHO) for use in drinking water cisterns to combat malaria. Most of its uses are to combat insects where the adult is the pest, including mosquitoes, several fly species, and fleas. Two very similar products, hydroprene and kinoprene, are used for controlling species such as cockroaches and white flies. Methoprene was registered with the EPA in 1975. Virtually no reports of resistance have been filed. A more recent type of IGR is the ecdysone agonist tebufenozide (MIMIC), which is used in forestry and other applications for control of caterpillars, which are far more sensitive to its hormonal effects than other insect orders.

Environmental Harm

Effects on Nontarget Species

Some insecticides kill or harm other creatures in addition to those they are intended to kill. For example, birds may be poisoned when they eat food that was recently sprayed with insecticides or when they mistake an insecticide granule on the ground for food and eat it. Sprayed insecticide may drift from the area to which it is applied and into wildlife areas, especially when it is sprayed aerially.

DDT

The development of DDT was motivated by desire to replace more dangerous or less effective alternatives. DDT was introduced to replace lead and arsenic-based compounds, which were in widespread use in the early 1940s.

DDT was brought to public attention by Rachel Carson's book *Silent Spring*. One side-effect of DDT is to reduce the thickness of shells on the eggs of predatory birds. The shells sometimes become too thin to be viable, reducing bird populations. This occurs with DDT and related compounds due to the process of bioaccumulation, wherein the chemical, due to its stability and fat solubility, accumulates in organisms' fatty tissues. Also, DDT may biomagnify, which causes progressively higher concentrations in the body fat of animals farther up the food chain. The near-worldwide ban on agricultural use of DDT and related chemicals has allowed some of these birds, such as the peregrine falcon, to recover in recent years. A number of organochlorine pesticides have been banned from most uses worldwide. Globally they are controlled via the Stockholm Convention on persistent organic pollutants. These include: aldrin, chlordane, DDT, dieldrin, endrin, heptachlor, mirex and toxaphene.

Pollinator Decline

Insecticides can kill bees and may be a cause of pollinator decline, the loss of bees that pollinate plants, and colony collapse disorder (CCD), in which worker bees from a beehive or Western honey bee colony abruptly disappear. Loss of pollinators means a reduction in crop yields. Sublethal doses of insecticides (i.e. imidacloprid and other neonicotinoids) affect bee foraging behavior. However, research into the causes of CCD was inconclusive as of June 2007.

Bird Decline

Besides the effects of direct consumption of insecticides, populations of insectivorous birds decline due to the collapse of their prey populations. Spraying of especially wheat and corn in Europe is believed to have caused an 80 per cent decline in flying insects, which in turn has reduced local bird populations by a third to two thirds.

Alternatives

Instead of using chemical insecticides to avoid crop damage caused by insects, there are many alternative options available now that can protect farmers from major economic losses. Some of them are:

- Breeding crops resistant, or at least less susceptible, to pest attacks.
- Releasing predators, parasitoids, or pathogens to control pest populations as a form of biological control.
- Chemical control like releasing pheromones into the field to confuse the insects into not being able to find mates and reproduce.
- Integrated Pest Management- using multiple techniques in tandem to achieve optimal results.
- Push-pull technique- intercropping with a "push" crop that repels the pest, and planting a "pull" crop on the boundary that attracts and traps it.

BIOPESTICIDE

Biopesticide is a contraction of 'biological pesticide', include several types of pest management intervention: through predatory, parasitic, or chemical relationships. The term has been associated historically with [biological control] – and by implication – the manipulation of living organisms. Regulatory positions can be influenced by public perceptions, thus:

- In the EU, biopesticides have been defined as "a form of pesticide based on micro-organisms or natural products".
- The US EPA states that they "include naturally occurring substances that control pests (biochemical pesticides), microorganisms that control pests (microbial pesticides), and pesticidal substances produced by plants containing added genetic material (plant-incorporated protectants) or PIPs".

They are obtained from organisms including plants, bacteria and other microbes, fungi, nematodes, *etc.* They are often important components of integrated pest management (IPM) programmes, and have received much practical attention as substitutes to synthetic chemical plant protection products (PPPs).

Types

Biopesticides can be classified into these classes:

- Microbial pesticides which consist of bacteria, entomopathogenic fungi or viruses (and sometimes includes the metabolites that bacteria or fungi produce). Entomopathogenic nematodes are also often classed as microbial pesticides, even though they are multi-cellular.

- Bio-derived chemicals. Four groups are in commercial use: pyrethrum, rotenone, neem oil, and various essential oils are naturally occurring substances that control (or monitor in the case of pheromones) pests and microbial diseases.
- Plant-incorporated protectants (PIPs) have genetic material from other species incorporated into their genetic material (i.e. GM crops). Their use is controversial, especially in many European countries.
- RNAi pesticides, some of which are topical and some of which are absorbed by the crop.

Biopesticides have usually no known function in photosynthesis, growth or other basic aspects of plant physiology. Instead, they are active against biological pests. Many chemical compounds have been identified that are produced by plants to protect them from pests so they are called antifeedants. These materials are biodegradable and renewable alternatives, which can be economical for practical use. Organic farming systems embraces this approach to pest control.

RNA

RNA interference is under study for possible use as a spray-on insecticide by multiple companies, including Monsanto, Syngenta, and Bayer. Such sprays do not modify the genome of the target plant. The RNA could be modified to maintain its effectiveness as target species evolve tolerance to the original. RNA is a relatively fragile molecule that generally degrades within days or weeks of application. Monsanto estimated costs to be on the order of $5/acre.

RNAi has been used to target weeds that tolerate Monsanto's Roundup herbicide. RNAi mixed with a silicone surfactant that let the RNA molecules enter air-exchange holes in the plant's surface that disrupted the gene for tolerance, affecting it long enough to let the herbicide work. This strategy would allow the continued use of glyphosate-based herbicides, but would not per se assist a herbicide rotation strategy that relied on alternating Roundup with others.

They can be made with enough precision to kill some insect species, while not harming others. Monsanto is also developing an RNA spray to kill potato beetles One challenge is to make it linger on the plant for a week, even if it's raining. The Potato beetle has become resistant to more than 60 conventional insecticides.

Monsanto lobbied the U.S. EPA to exempt RNAi pesticide products from any specific regulations (beyond those that apply to all pesticides) and be exempted from rodent toxicity, allergenicity and residual environmental testing. In 2014 an EPA advisory group found little evidence of a risk to people from eating RNA.

However, in 2012, the Australian Safe Food Foundation alleged that the RNA trigger designed to change wheat's starch content might interfere with the gene for a human liver enzyme. Supporters countered that RNA does not appear to make it past human saliva or stomach acids. The US National Honey Bee Advisory Board told EPA that using RNAi would put natural systems at "the epitome of risk". The beekeepers cautioned that pollinators could be hurt by unintended effects and that the genomes of many insects are still unknown. Other unassessed risks include ecological (given the need for sustained presence for herbicide and other applications) and the possible for RNA drift across species boundaries.

Monsanto has invested in multiple companies for their RNA expertise, including Beeologics (for RNA that kills a parasitic mite that infests hives and for manufacturing technology) and Preceres (nanoparticle lipidoid coatings) and licensed technology from Alnylam and Tekmira. In 2012 Syngenta acquired Devgen, a European RNA partner. Startup Forrest Innovations is investigating RNAi as a solution to citrus greening disease that in 2014 caused 22 percent of oranges in Florida to fall off the trees.

Examples

Bacillus thuringiensis, a bacterial disease of Lepidoptera, Coleoptera and Diptera, is a well-known insecticide example. The toxin from B. thuringiensis (Bt toxin) has been incorporated directly into plants through the use of genetic engineering. The use of Bt Toxin is particularly controversial. Its manufacturers claim it has little effect on other organisms, and is more environmentally friendly than synthetic pesticides.

Other microbial control agents include products based on:

- Entomopathogenic fungi (e.G. Beauveria bassiana, isaria fumosorosea, lecanicillium and metarhizium spp.).
- Plant disease control agents: include trichoderma spp. And ampelomyces quisqualis (a hyper-parasite of grape powdery mildew); bacillus subtilis is also used to control plant pathogens.
- Beneficial nematodes attacking insect (e.G. Steinernema feltiae) or slug (e.G. Phasmarhabditis hermaphrodita) pests.
- Entomopathogenic viruses (e.G. Cydia pomonella granulovirus).
- Weeds and rodents have also been controlled with microbial agents.

Various naturally occurring materials, including fungal and plant extracts, have been described as biopesticides. Products in this category include:

- Insect pheromones and other semiochemicals.

- Fermentation products such as Spinosad (a macro-cyclic lactone).
- Chitosan: A plant in the presence of this product will naturally induce systemic resistance (ISR) to allow the plant to defend itself against disease, pathogens and pests.
- Biopesticides may include natural plant-derived products, which include alkaloids, terpenoids, phenolics and other secondary chemicals. Certain vegetable oils such as canola oil are known to have pesticidal properties. Products based on extracts of plants such as garlic have now been registered in the EU and elsewhere.

Applications

Biopesticides are biological or biologically-derived agents, that are usually applied in a manner similar to chemical pesticides, but achieve pest management in an environmentally friendly way. With all pest management products, but especially microbial agents, effective control requires appropriate formulation and application.

Biopesticides for use against crop diseases have already established themselves on a variety of crops. For example, biopesticides already play an important role in controlling downy mildew diseases. Their benefits include: a 0-Day Pre-Harvest Interval, the ability to use under moderate to severe disease pressure, and the ability to use as a tank mix or in a rotational program with other registered fungicides. Because some market studies estimate that as much as 20% of global fungicide sales are directed at downy mildew diseases, the integration of biofungicides into grape production has substantial benefits in terms of extending the useful life of other fungicides, especially those in the reduced-risk category.

A major growth area for biopesticides is in the area of seed treatments and soil amendments. Fungicidal and biofungicidal seed treatments are used to control soil borne fungal pathogens that cause seed rots, damping-off, root rot and seedling blights. They can also be used to control internal seed–borne fungal pathogens as well as fungal pathogens that are on the surface of the seed. Many biofungicidal products also show capacities to stimulate plant host defence and other physiological processes that can make treated crops more resistant to a variety of biotic and abiotic stresses.

Disadvantages

- High specificity: which may require an exact identification of the pest/pathogen and the use of multiple products to be used; although this can also be an advantage in that the biopesticide is less likely to harm species other than the target.
- Often slow speed of action (thus making them unsuitable if a pest outbreak is an immediate threat to a crop).

- Often variable efficacy due to the influences of various biotic and abiotic factors (since some biopesticides are living organisms, which bring about pest/pathogen control by multiplying within or nearby the target pest/pathogen).
- Living organisms evolve and increase their resistance to biological, chemical, physical or any other form of control. If the target population is not exterminated or rendered incapable of reproduction, the surviving population can acquire a tolerance of whatever pressures are brought to bear, resulting in an evolutionary arms race.
- Unintended consequences: Studies have found broad spectrum biopesticides have lethal and nonlethal risks for non-target native pollinators such as *Melipona quadrifasciata* in Brazil.

BIRD NETTING

Bird netting or anti-bird netting is a form of bird pest control. It is a net used to prevent birds from reaching certain areas.

Bird protection netting comes in a variety of shapes and forms, The most common is a small mesh (1 or 2 cm squares) either extruded and bi-oriented polypropylene or woven polyethylene.

The color most used is black (as the carbon black UV inhibitor offers the best protection against solar rays), but also bird netting may be available in other colors like white (usually white netting is woven or knitted and has an even smaller mesh size as it will serve as a double purpose anti-hail net for the protection of fruits during summer hail storms or late spring during flowering) or green (usually used in home gardening and mostly sold at retail outlets for the DIY farmers).

Professional anti-bird netting comes in jumbo rolls that will offer considerable savings to the farmers or aquaculturalist. Retail chains and local stores will offer smaller packages that fit the backyard gardener's needs.

Usages

Crop Protection

Bird nets are used to prevent bird damage of vegetable and fruit crops as well as seedlings. Frugivore birds and bats can cause great damages to farmers as they tend to peck one fruit, then go to another, therefore ruining a large percentage of otherwise commercially valuable production. Once even a small portion is bitten off, that fruit cannot be sold and if harvested (even if there is no bacteria or virus brought by the frugivore) will rot or ferment, damaging the rest of the harvested case. Bird protection netting is applied directly on the stand-alone trees or espaliers like peaches, pears, apples, grapes, or on the side ventilation windows of growing tunnels as in the case of berries like strawberries, raspberries, blueberries, and cranberries.

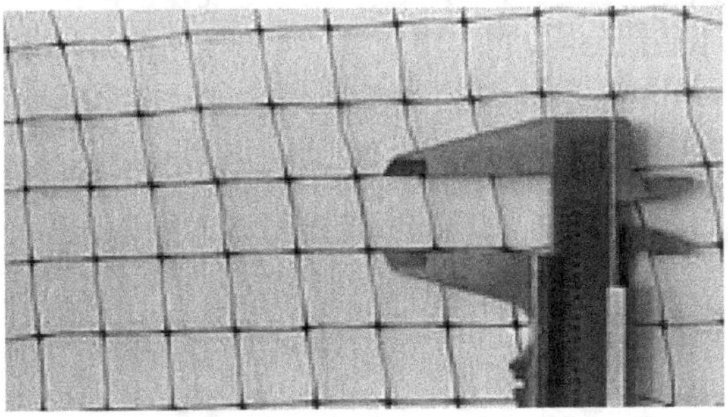

Anti-bird netting size.

Fish Protection

Bird netting may be used to protect fisheries and fish wildlife reserves from predator birds. Also in aquaculture (like shrimp and tilapia farms to mention a few), growers need to protect their work and fish crops from marauding birds. These type of birds have usually a larger wing span (seagulls, pelicans, herons, cormorants etc.) and a larger mesh size (with individual strands being more resistant as it will be istalled on a cable system crossing the growing ponds). These netting are usually white as to be very visible for the large sea birds will be deterred by the sight of a barrier to their diving into the ponds.

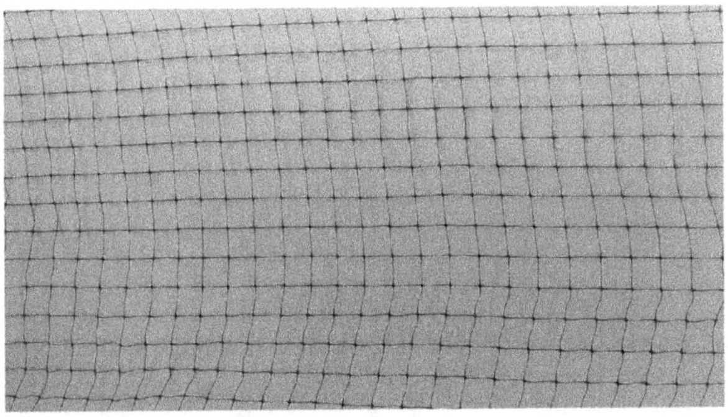

Anti-bird-netting, malla-antipajaros example.

Building Protection

Bird netting is one of the most effective and long lasting ways of bird proofing buildings and other structures against all urban bird species. It provides a discreet and impenetrable barrier that protects premises without harming the birds. Bird netting can be particularly effective for large open areas such as roofs and loading bays. Design considerations include the type and material of the fixings utilized and the bird species requiring exclusion.

Mining Ponds

Miners will use chemical agents to extract minerals or metals from crushed rocks. These harmful chemical agents must be confined from volatiles, especially migratory species. In the United States EPA mandates that such cyanide ponds be covered at all times to prevent loss of wildlife.

BIRD SCARER

A traditional scarecrow.

Bird scarers are a number of devices designed to scare birds, usually employed by farmers to dissuade birds from eating recently planted arable crops.

They are also used on airfields to prevent birds accumulating near runways and causing a potential hazard to aircraft.

Visual Scarers

Scarecrow

One of the oldest designs of bird scarer is the scarecrow which is in the shape of a human figure. The scarecrow idea has been built upon numerous times, and not all visual scare devices are shaped like humans. The "Flashman Birdscarer," Iridescent tape, "TerrorEyes" balloons, and other visual deterrents are all built on the idea of visually scaring birds. This method doesn't work so well with all species, considering that some species frequently perch on scarecrows.

Hawk Kite

Many species of bird are also naturally afraid of predators such as birds of prey. "Hawk kites" are designed to fly from poles in the wind and hover above the field to be protected. They are shaped to match the silhouette of a bird of prey.

A stationary modelled owl used as a bird scarer at Lysaker, Norway.

Helikites

The Helikite bird scarer is a lighter-than-air combination of a helium balloon and a kite. Helikites fly up to 200vft in the air with or without wind. Although they do not look like hawks, they fly and hover high in the sky behaving like birds of prey. Helikites successfully exploit bird pests' instinctive fear of hawks and can reliably protect large areas of farmland.

Lasers

The use of lasers can be an effective method of bird scaring, although there is some evidence to suggest some birds are "laser-resistant". As the effectiveness of the laser decreases with increasing light levels, it is likely to be most effective at dawn and dusk. Although some lasers prove to be effective during daylight hours.

The method relies on birds being startled by the strong contrast between the ambient light and the laser beam. During low light conditions this technique is very selective and can be attuned to frequencies and wavelengths that individual bird species don't like, but at night the light beam is visible over a large distance and can cause widespread (non-species specific) disturbance. Lasers use can be limited due to safety concerns of the beam and some nations have laws which prohibit lasers above a certain power from being used. Lasers should be looked at as an additional scaring system to add to a wildlife management program.

Manually operated laser torches and automated laser bird deterrent robots that move the laser automatically towards the birds are available on the market.

Dead Birds

The use of model or actual dead birds is used to signal danger to others. Initially, birds often approach the corpse but usually leave when they see the unnatural position of the bird. This approach has been frequently used in attempts to deter gulls from airports.

Pheasant feed sacks often have an image of an owl with large eyes so that when empty they can be strung up to scare predators.

Balloons

An example of a visual bird-scare balloon.

Balloons are an inexpensive deterrent. However, this method relies on the movement of balloons, which is something that birds can become used to. The addition of eye illustrations on the balloons has been shown to increase this method's effectiveness as it combats the birds' ability to adapt. Commercially available "scare-eye" balloons have holographic eyes that follow birds wherever they go. The long-term effectiveness of this method can be increased by periodically moving the placement of the scare devices.

In the United Kingdom the use of balloons is subject to approval from the Civil Aviation Authority, especially around airfields.

Auditory Scarers

Audible bird scarers use noise stimuli that makes birds uncomfortable. However, once birds realize these pose no real threats, they can easily become habituated to sounds that seemed initially frightening. If just being placed in situ and left, audible bird scarers can easily become ineffective bird control solutions, however when managed on an ongoing basis or used as part of a greater bird deterrent system, sound methods can deliver quality results.

Sōzu (Shishi Odoshi)

One very old design is the Japanese sōzu, known metonymically as a shishi odoshi (although the term *shishi odoshi* properly refers to any method of scaring wild animals, including the Western scarecrow). Instead of using a visual method to distract pests, as the scarecrow does, it uses the sound of a heavy pipe repeatedly and rhythmically hitting a rock, using water as a timing device. The sōzu is also used in Japanese popular culture to denote inordinate amounts of wealth, combined with a traditional sensibility: by design, the shishi odoshi uses copious amounts of water, meaning either a very high water bill, or that it is situated on high-value land with a stream or river running through it.

Propane Cannons

A typical propane gas gun bird scarer.

Propane scare cannons are one of the most common types of bird scarer available in Europe and America. It is a propane-powered gas gun which produces a periodic explosion. The audible bang can reach very loud volumes, in excess of 150 decibels, causing a flight reaction in birds. The similarity between a scare cannon and a 12 gauge shotgun is thought to cause a startle/fear reaction, although it is also effective against birds that have not been exposed to hunting pressure.

Birds can become habituated to the sound of regular cannon detonations, especially if it does not vary in its magnitude, pitch, or time interval. However, regularly moving the cannon, utilizing on-demand firing options, including radio control, and combining cannons with other methods of deterrents can prevent habituation.

Propane scare cannons are very loud and can be disruptive to people living nearby. One study found that restricting cannon use to only hours when birds are active and incorporating better bird damage plans drastically reduced the number of complaints from neighbours.

Electronic Repellers

Sonic bird repellers are not effective; the birds quickly acclimate to them. Usually consisting of a central unit and several speakers, the system emits digitally recorded distress calls of birds, and, in some cases, calls of predators of the target species. Some emitters randomize pitch, magnitude, time interval, sound sequence and other factors in an attempt to prevent birds from getting used to them. Many of the sounds produced are regarded as annoying to people.

Ultrasonic Scarers

Ultrasonic devices are static sound-emitting bird deterrents, which, in theory, will annoy birds to keep them away from enclosed or semi-enclosed areas. Ultrasonic scarers are not harmful to birds, however, there is debate around birds' ability to hear these frequencies at loud enough decibels. Birds are believed to have similar hearing to humans, with studies showing birds do not hear on an ultrasonic level, meaning that ultrasonic scarers often have little or no effect in deterring birds.

Cartridge Scarers

Cartridge scarers include a wide variety of noise-producing cartridges usually fired from rockets or rope bangers, or on aerodromes from modified pistols or shotguns, which produce a loud bang and emit flashes of light. They include shellcrackers, screamer shells and whistling projectiles, exploding projectiles, bird bangers and flares. Bird banger cartridges commonly use a low explosive known as flash powder.

Cartridges are projected from a shotgun with a range of 45–90 metres (148–295 ft), or pistols with a range of approximately 25 metres (82 ft), before exploding. Bird scaring cartridges can produce noise levels of up to 160 dB at varying ranges but in some countries both the cartridges and the gun require a firearms certificate.

Pyrotechnics have proved effective in dispersing birds at airports, landfill sites, agricultural crops and aquaculture facilities. At airports in the United Kingdom, shellcrackers fired from a modified pistol are the most common means of dispersing birds, as they allow the bird controller to have some directional control over birds in flight, so they can be steered away from runways.

However, as with all similar noises, there is a high risk of birds becoming used to any pyrotechnics or cartridge explosions.

Benign Acoustic Deterrence

In 2013, Dr. John Swaddle and Dr. Mark Hinders at the College of William and Mary created a new method of deterring birds using benign sounds projected by conventional and directional (parametric) speakers. The initial objectives of the technology were to displace problematic birds from airfields to reduce bird strike risks, minimize agricultural losses due to pest bird foraging, displace nuisance birds that cause extensive repair and chronic clean-up costs, and reduce bird mortality from flying into man-made structures. The sounds, referred to as a "Sonic Net," do not have to be loud and are a combination of wave forms—collectively called "colored noise"—forming non-constructive and constructive interference with how birds talk to each other.

Technically, the Sonic Net technology is not a bird scarer, but discourages birds from flying into or spending time in the target area. The impact on the birds is similar to talking in a crowded room, and since they cannot understand each other they go somewhere else. Early tests at an aviary and initial field trials at a landfill and airfield indicate that the technology is effective and that birds do not habituate to the sound. The provisional and full patents were filed in 2013 and 2014 respectively, with further research and commercialization of the technology being ongoing.

Other

Humans

Historically, humans have been employed to scare birds from crops, using a variety of deterrents including throwing stones, flashing with mirrors, or operating noise devices. This is only cost-effective where the cost of labour is sufficiently low relative to agricultural profit margins. In Victorian England, children were employed to do this job. In the Huleh Valley in Northern Israel, rural labourers, primarily Arabs and Druze, use USVs and multiple devices to keep common cranes off high-value crops.

Dogs

The control of birds and other wildlife such as deer through harassment by trained border collies has been used at aerodromes, golf courses and agricultural land. The dogs represent an actual threat, and so elicit flight reactions. Habituation is unlikely as they can continually pursue and change their behaviour. Border collies are used as they are working dogs bred to herd animals and to avoid attack, and they respond well to whistle and verbal commands. A single border collie and its handler can keep an area of approximately 50 square kilometres (19.3 square miles, 4998.7 hectares, or 12,179.2 acres) free of larger birds and wildlife. However, although they are effective at deterring ground foraging birds such as waders and wildfowl, they are not so useful for species that spend most of their time flying or perching, such as raptors and swallows.

In 1999, Southwest Florida International Airport became the first commercial airport in the world to employ a border collie in an airfield wildlife control programme. After the use of the collie, numbers and species of birds on the airport declined and most birds that remained congregated in a drainage ditch away from the runway. The number of bird strikes dropped to zero compared to 13 for the same period the previous year. Several other airports and airbases have now started similar programmes.

At Dover Air Force Base, Delaware, bird strike damage to aircraft caused by birds has been reduced from an average of US$600,000 per year for the proceeding two years to US$24,000 per year after the initiation of a bird control programme that included the use of border collies.

Predators

Using predators as a natural bird deterrent has become a recommended form of controlling bird infestations. Specially selected species are trained to deal with working in un-natural environments with distractions and dangers they would not usually encounter.

The success of this method of bird control is based on the fact that many birds have a natural fear of falcons and hawks as predators, so their presence in the area encourages problem species to disperse. The natural reaction of most prey species is to form a flock and attempt to fly above the falcon. If this fails, they will attempt to fly for cover and leave the area.

Harris's hawk.

Radio Controlled Aircraft

Radio-controlled model aircraft have been used to scare or 'haze' bird pests since the early 1980s, mainly over airfields, but have also been used over agricultural areas, fisheries and landfill sites. This method has been shown to be very effective and birds habituate more slowly to a treatment in which they are being actively hazed. At Whiteman Air Force Base, Missouri, balsa wood radio-controlled aircraft are one of the primary bird harassment methods used to keep the airfield clear of raptors and other large birds, and they have also proved effective at dispersing the base's redwing blackbird roost.

Fireworks

Fireworks can also be used as bird scarers, and some jurisdictions issue special licences for agricultural fireworks. This practice has been criticised as a loophole for the sale of consumer fireworks. Again, the loud bangs can also irritate people living on nearby properties.

Combination Scarers

These combine multiple deterrents, such as using a pop up scarecrow combined with a gas gun, which in turn activates the distress call of a bird. These combination scarers are often managed by computers and synchronised across an area via the use of radio links. This synchronisation becomes more effective if there is some kind of detection system involved such as bird detecting radar.

INTEGRATED PEST MANAGEMENT

Integrated pest management (IPM), also known as integrated pest control (IPC) is a broad-based approach that integrates practices for economic control of pests. IPM aims to suppress pest populations below the economic injury level (EIL). The UN's Food and Agriculture Organization defines IPM as "the careful consideration of all available pest control techniques and subsequent integration of appropriate measures that discourage the development of pest populations and keep pesticides and other interventions to levels that are economically justified and reduce or minimize risks to human health and the environment. IPM emphasizes the growth of a healthy crop with the least possible disruption to agro-ecosystems and encourages natural pest control mechanisms." Entomologists and ecologists have urged the adoption of IPM pest control since the 1970s. IPM allows for safer pest control.

The introduction and spread of invasive species can also be managed with IPM by reducing risks while maximizing benefits and reducing costs.

Shortly after World War II, when synthetic insecticides became widely available, entomologists in California developed the concept of "supervised insect control". Around the same time, entomologists in the US Cotton Belt were advocating a similar approach. Under this scheme, insect control was "supervised" by qualified entomologists and insecticide applications were based on conclusions reached from periodic monitoring of pest and natural-enemy populations. This was viewed as an alternative to calendar-based programs. Supervised control was based on knowledge of the ecology and analysis of projected trends in pest and natural-enemy populations.

Supervised control formed much of the conceptual basis for the "integrated control" that University of California entomologists articulated in the 1950s. Integrated control sought to identify the best mix of chemical and biological controls for a given insect pest. Chemical insecticides were to be used in the manner least disruptive to biological control. The term "integrated" was thus synonymous with "compatible." Chemical controls were to be applied only after regular monitoring indicated that a pest population had reached a level (the economic threshold) that required treatment to prevent the population from reaching a level (the economic injury level) at which economic losses would exceed the cost of the control measures.

IPM extended the concept of integrated control to all classes of pests and was expanded to include all tactics. Controls such as pesticides were to be applied as in integrated control, but these now had to be compatible with tactics for all classes of pests. Other tactics, such as host-plant resistance and cultural manipulations, became part of the IPM framework. IPM combined entomologists, plant pathologists, nematologists and weed scientists.

In the United States, IPM was formulated into national policy in February 1972 when President Richard Nixon directed federal agencies to take steps to advance the application of IPM in all relevant sectors. In 1979, President Jimmy Carter established an interagency IPM Coordinating Committee to ensure development and implementation of IPM practices.

Perry Adkisson and Ray F. Smith received the 1997 World Food Prize for encouraging the use of IPM.

Applications

IPM is used in agriculture, horticulture, forestry, human habitations, preventive conservation and general pest control, including structural pest management, turf pest management and ornamental pest management.

Principles

An American IPM system is designed around six basic components:

- Acceptable pest levels—The emphasis is on control, not eradication. IPM holds that wiping out an entire pest population is often impossible, and the attempt can be expensive and unsafe. IPM programmes first work to establish acceptable pest levels, called action thresholds, and apply controls if those thresholds are crossed. These thresholds are pest and site specific, meaning that it may be acceptable at one site to have a weed such as white clover, but not at another site. Allowing a pest population to survive at a reasonable threshold reduces selection pressure. This lowers the rate at which a pest develops resistance to a control, because if almost all pests are killed then those that have resistance will provide the genetic basis of the future population. Retaining a significant number of unresistant specimens dilutes the prevalence of any resistant genes that appear. Similarly, the repeated use of a single class of controls will create pest populations that are more resistant to that class, whereas alternating among classes helps prevent this.

- Preventive cultural practices—Selecting varieties best for local growing conditions and maintaining healthy crops is the first line of defense. Plant quarantine and 'cultural techniques' such as crop sanitation are next, e.g., removal of diseased plants, and cleaning pruning shears to prevent spread of infections. Beneficial fungi and bacteria are added to the potting media of horticultural crops vulnerable to root diseases, greatly reducing the need for fungicides.

- Monitoring—Regular observation is critically important. Observation is broken into inspection and identification. Visual inspection, insect and spore traps, and other methods are used to monitor pest levels. Record-keeping is essential, as is a thorough knowledge of target pest behavior and reproductive cycles. Since insects are cold-blooded, their physical development is dependent on area temperatures. Many insects have had their development cycles modeled in terms of degree-days. The degree days of an environment determines the optimal time for a specific insect outbreak. Plant pathogens follow similar patterns of response to weather and season.
- Mechanical controls—Should a pest reach an unacceptable level, mechanical methods are the first options. They include simple hand-picking, barriers, traps, vacuuming and tillage to disrupt breeding.
- Biological controls—Natural biological processes and materials can provide control, with acceptable environmental impact, and often at lower cost. The main approach is to promote beneficial insects that eat or parasitize target pests. Biological insecticides, derived from naturally occurring microorganisms (e.g.—Bt, entomopathogenic fungi and entomopathogenic nematodes), also fall in this category. Further 'biology-based' or 'ecological' techniques are under evaluation.
- Responsible use—Synthetic pesticides are used as required and often only at specific times in a pest's life cycle. Many newer pesticides are derived from plants or naturally occurring substances (e.g.—nicotine, pyrethrum and insect juvenile hormone analogues), but the toxophore or active component may be altered to provide increased biological activity or stability. Applications of pesticides must reach their intended targets. Matching the application technique to the crop, the pest, and the pesticide is critical. The use of low-volume spray equipment reduces overall pesticide use and labor cost.

An IPM regime can be simple or sophisticated. Historically, the main focus of IPM programmes was on agricultural insect pests. Although originally developed for agricultural pest management, IPM programmes are now developed to encompass diseases, weeds and other pests that interfere with management objectives for sites such as residential and commercial structures, lawn and turf areas, and home and community gardens.

Process

IPM is the selection and use of pest control actions that will ensure favourable economic condition, ecological and social consequences and is applicable to most agricultural, public health and amenity pest management situations. The IPM process starts with monitoring, which includes inspection and identification, followed by the establishment of economic injury levels. The economic injury levels set the economic threshold level. That is the point when pest damage (and the benefits of treating the pest) exceed the cost of treatment. This can also be an action threshold level for determining an unacceptable level that is not tied to economic injury. Action thresholds are more common in structural pest management and economic injury levels in classic agricultural pest management. An example of an action threshold is one fly in a hospital operating room is not acceptable, but one fly in a pet kennel would be acceptable. Once a threshold has been crossed by the pest population action steps need to be taken to reduce and control the pest. Integrated pest management employs a variety of actions including cultural controls such as physical barriers, biological controls such as adding and conserving natural predators and enemies of the pest, and finally chemical controls or pesticides. Reliance on knowledge, experience, observation and integration of multiple techniques makes IPM appropriate for organic farming (excluding synthetic pesticides). These may or may not include materials listed on the Organic Materials Review Institute (OMRI) Although the pesticides and particularly insecticides used in organic farming and organic gardening are generally safer than synthetic pesticides, they are not always more safe or environmentally friendly than synthetic pesticides and can cause harm. For conventional farms IPM can reduce human and environmental exposure to hazardous chemicals, and potentially lower overall costs.

Risk assessment usually includes four issues: 1) characterization of biological control agents, 2) health risks, 3) environmental risks and 4) efficacy.

Mistaken identification of a pest may result in ineffective actions. E.g., plant damage due to over-watering could be mistaken for fungal infection, since many fungal and viral infections arise under moist conditions.

Monitoring begins immediately, before the pest's activity becomes significant. Monitoring of agricultural pests includes tracking soil/planting media fertility and water quality. Overall plant health and resistance to pests is greatly influenced by pH, alkalinity, of dissolved mineral and oxygen reduction potential. Many diseases are waterborne, spread directly by irrigation water and indirectly by splashing.

Once the pest is known, knowledge of its lifecycle provides the optimal intervention points. For example, weeds reproducing from last year's seed can be prevented with mulches and pre-emergent herbicide.

Pest-tolerant crops such as soybeans may not warrant interventions unless the pests are numerous or rapidly increasing. Intervention is warranted if the expected cost of damage by the pest is more than the cost of control. Health hazards may require intervention that is not warranted by economic considerations.

Specific sites may also have varying requirements. E.g., white clover may be acceptable on the sides of a tee box on a golf course, but unacceptable in the fairway where it could confuse the field of play.

Possible interventions include mechanical/physical, cultural, biological and chemical. Mechanical/physical controls include picking pests off plants, or using netting or other material to exclude pests such as birds from grapes or rodents from structures. Cultural controls include keeping an area free of conducive conditions by removing waste or diseased plants, flooding, sanding, and the use of disease-resistant crop varieties. Biological controls are numerous. They include: conservation of natural predators or augmentation of natural predators, sterile insect technique (SIT).

Augmentation, inoculative release and inundative release are different methods of biological control that affect the target pest in different ways. Augmentative control includes the periodic introduction of predators. With inundative release, predators are collected, mass-reared and periodically released in large numbers into the pest area. This is used for an immediate reduction in host populations, generally for annual crops, but is not suitable for long run use. With inoculative release a limited number of beneficial organisms are introduced at the start of the growing season. This strategy offers long term control as the organism's progeny affect pest populations throughout the season and is common in orchards. With seasonal inoculative release the beneficials are collected, mass-reared and released seasonally to maintain the beneficial population. This is commonly used in greenhouses. In America and other western countries, inundative releases are predominant, while Asia and the eastern Europe more commonly use inoculation and occasional introductions.

The sterile insect technique (SIT) is an area-wide IPM program that introduces sterile male pests into the pest population to trick females into (unsuccessful) breeding encounters, providing a form of birth control and reducing reproduction rates. The biological controls mentioned above only appropriate in extreme cases, because in the introduction of new species, or supplementation of naturally occurring species can have detrimental ecosystem effects. Biological controls can be used to stop invasive species or pests, but they can become an introduction path for new pests.

Chemical controls include horticultural oils or the application of insecticides and herbicides. A green pest management IPM program uses pesticides derived from plants, such as botanicals, or other naturally occurring materials.

Pesticides can be classified by their modes of action. Rotating among materials with different modes of action minimizes pest resistance.

Evaluation is the process of assessing whether the intervention was effective, whether it produced unacceptable side effects, whether to continue, revise or abandon the program.

WEED CONTROL

Weed control is the botanical component of pest control, which attempts to stop weeds, especially noxious weeds, from competing with desired flora and fauna including domesticated plants and livestock, and in natural settings preventing non native species competing with native species.

Weed control is important in agriculture. Methods include hand cultivation with hoes, powered cultivation with cultivators, smothering with mulch, lethal wilting with high heat, burning, and chemical control with herbicides (weed killers).

The definition of a weed is context-dependent, and can be taken to mean any plant not wanted for any reason.

Weeds compete with productive crops or pasture, ultimately converting productive land into unusable scrub. Weeds can be poisonous, distasteful, produce burrs, thorns or otherwise interfere with the use and management of desirable plants by contaminating harvests or interfering with livestock.

Weeds compete with crops for space, nutrients, water and light. Smaller, slower growing seedlings are more susceptible than those that are larger and more vigorous. Onions are one of the most vulnerable, because they are slow to germinate and produce slender, upright stems. By contrast broad beans produce large seedlings and suffer far fewer effects other than during periods of water shortage at the crucial time when the pods are filling out. Transplanted crops raised in sterile soil or potting compost gain a head start over germinating weeds.

Weeds also vary in their competitive abilities and according to conditions and season. Tall-growing vigorous weeds such as fat hen (*Chenopodium album*) can have the most pronounced effects on adjacent crops, although seedlings of fat hen that appear in late summer produce only small plants. Chickweed (*Stellaria media*), a low growing plant, can happily co-exist with a tall crop during the summer, but plants that have overwintered will grow rapidly in early spring and may swamp crops such as onions or spring greens.

The presence of weeds does not necessarily mean that they are damaging a crop, especially during the early growth stages when both weeds and crops can grow without interference. However, as growth proceeds they each begin to require greater amounts of water and nutrients. Estimates suggest that weed and crop can co-exist harmoniously for around three weeks before competition becomes significant. One study found that after competition had started, the final yield of onion bulbs was reduced at almost 4% per day.

Perennial weeds with bulbils, such as lesser celandine and oxalis, or with persistent underground stems such as couch grass (*Agropyron repens*) or creeping buttercup (*Ranunculus repens*) store reserves of food, and are thus able to grow faster and with more vigour than their annual counterparts. Some perennials such as couch grass exude allelopathic chemicals that inhibit the growth of other nearby plants.

Weeds can also host pests and diseases that can spread to cultivated crops. Charlock and Shepherd's purse may carry clubroot, eelworm can be harboured by chickweed, fat hen and shepherd's purse, while the cucumber mosaic virus, which can devastate the cucurbit family, is carried by a range of different weeds including chickweed and groundsel.

Insect pests often do not attack weeds. However pests such as cutworms may first attack weeds then move on to cultivated crops.

Some plants are considered weeds by some farmers and crops by others. Charlock, a common weed in the southeastern US, are weeds according to row crop growers, but are valued by beekeepers, who seek out places where it blooms all winter, thus providing pollen for honeybees and other pollinators. Its bloom resists all but a very hard freeze, and recovers once the freeze ends.

Weed Propagation

Seeds

Annual and biennial weeds such as chickweed, annual meadow grass, shepherd's purse, groundsel, fat hen, cleaver, speedwell and hairy bittercress propagate themselves by seeding. Many produce huge numbers of seed several times a season, some all year round. Groundsel can produce 1000 seed, and can continue right through a mild winter, whilst Scentless Mayweed produces over 30,000 seeds per plant. Not all of these will germinate at once, but over several seasons, lying dormant in the soil sometimes for years until exposed to light. Poppy seed can survive 80–100 years, dock 50 or more. There can be many thousands of seeds in a square foot or square metre of ground, thus any soil disturbance will produce a flush of fresh weed seedlings.

Subsurface/Surface

The most persistent perennials spread by underground creeping rhizomes that can regrow from a tiny fragment. These include couch grass, bindweed, ground elder, nettles, rosebay willow herb, Japanese knotweed, horsetail and bracken, as well as creeping thistle, whose tap roots can put out lateral roots. Other perennials put out runners that spread along the soil surface. As they creep they set down roots, enabling them to colonise bare ground with great rapidity. These include creeping buttercup and ground ivy. Yet another group of perennials propagate by stolons- stems that arch back into the ground to reroot. The most familiar of these is the bramble.

Methods

Pesticide-free thermic weed control with a weed burner on a potato field in Dithmarschen, Germany.

Weed control plans typically consist of many methods which are divided into biological, chemical, cultural, and physical/mechanical control.

Physical/Mechanical Methods

Coverings

In domestic gardens, methods of weed control include covering an area of ground with a material that creates a hostile environment for weed growth, known as a *weed mat*.

Several layers of wet newspaper prevent light from reaching plants beneath, which kills them. Daily saturating the newspaper with water plant decomposition. After several weeks, all germinating weed seeds are dead.

In the case of black plastic, the greenhouse effect kills the plants. Although the black plastic sheet is effective at preventing weeds that it covers, it is difficult to achieve complete coverage. Eradicating persistent perennials may require the sheets to be left in place for at least two seasons.

Some plants are said to produce root exudates that suppress herbaceous weeds. *Tagetes minuta* is claimed to be effective against couch and ground elder, whilst a border of comfrey is also said to act as a barrier against the invasion of some weeds including couch. A 5–10 centimetres (2.0–3.9 in) layer of wood chip mulch prevents most weeds from sprouting.

Gravel can serve as an inorganic mulch.

Tools used for amateur weeding include spades and gloves.

Irrigation is sometimes used as a weed control measure such as in the case of paddy fields to kill any plant other than the water-tolerant rice crop.

Manual Removal

Many gardeners still remove weeds by manually pulling them out of the ground, making sure to include the roots that would otherwise allow them to resprout.

Weeds are removed manually in large parts of India.

Hoeing off weed leaves and stems as soon as they appear can eventually weaken and kill perennials, although this will require persistence in the case of plants such as bindweed. Nettle infestations can be tackled by cutting back at least three times a year, repeated over a three-year period. Bramble can be dealt with in a similar way.

A highly successful, mostly manual, removal programme of weed control in natural bush land has been the control of sea spurge by Sea Spurge Remote Area Teams in Tasmania.

Tillage

Ploughing includes tilling of soil, intercultural ploughing and summer ploughing. Ploughing uproots weeds, causing them to die. In summer ploughing is done during deep summers. Summer ploughing also helps in killing pests.

Mechanical tilling can remove weeds around crop plants at various points in the growing process.

An Aquamog can be used to remove weeds covering a body of water.

Thermal

Several thermal methods can control weeds.

Flame weeders use a flame several centimeters away from the weeds to give them a sudden and severe heating. The goal of flame weeding is not necessarily burning the plant, but rather causing a lethal wilting by denaturing proteins in the weed. Similarly, hot air weeders can heat up the seeds to the point of destroying them. Flame weeders can be combined with techniques such as stale seedbeds (preparing and watering the seedbed early, then killing the nascent crop of weeds that springs up from it, then sowing the crop seeds) and preemergence flaming (doing a flame pass against weed seedlings after the sowing of the crop seeds but before those seedlings emerge from the soil—a span of time that can be days or weeks).

Hot foam (foamstream) causes the cell walls to rupture, killing the plant. Weed burners heat up soil quickly and destroy superficial parts of the plants. Weed seeds are often heat resistant and even react with an increase of growth on dry heat.

Since the 19th century soil steam sterilization has been used to clean weeds completely from soil. Several research results confirm the high effectiveness of humid heat against weeds and its seeds.

Soil solarization in some circumstances is very effective at eliminating weeds while maintaining grass. Planted grass tends to have a higher heat/humidity tolerance than unwanted weeds.

Seed Targeting

In 1998, the Australian Herbicide Resistance Initiative (AHRI), debuted. gathered fifteen scientists and technical staff members to conduct field surveys, collect seeds, test for resistance and study the biochemical and genetic mechanisms of resistance. A collaboration with DuPont led to a mandatory herbicide labeling program, in which each mode of action is clearly identified by a letter of the alphabet.

The key innovation of the AHRI approach has been to focus on weed seeds. Ryegrass seeds last only a few years in soil, so if farmers can prevent new seeds from arriving, the number of sprouts will shrink each year. Until the new approach farmers were unintentionally helping the seeds. Their combines loosen ryegrass seeds from their stalks and spread them over the fields. In the mid-1980s, a few farmers hitched covered trailers, called "chaff carts", behind their combines to catch the chaff and weed seeds. The collected material is then burned.

An alternative is to concentrate the seeds into a half-meter-wide strip called a windrow and burn the windrows after the harvest, destroying the seeds. Since 2003, windrow burning has been adopted by about 70% of farmers in Western Australia.

Yet another approach is the Harrington Seed Destructor, which is an adaptation of a coal pulverizing cage mill that uses steel bars whirling at up to 1500 rpm. It keeps all the organic material in the field and does not involve combustion, but kills 95% of seeds.

Cultural Methods

Stale Seed Bed

Another manual technique is the 'stale seed bed', which involves cultivating the soil, then leaving it fallow for a week or so. When the initial weeds sprout, the grower lightly hoes them away before planting the desired crop. However, even a freshly cleared bed is susceptible to airborne seed from elsewhere, as well as seed carried by passing animals on their fur, or from imported manure.

Buried Drip Irrigation

Buried drip irrigation involves burying drip tape in the subsurface near the planting bed, thereby limiting weeds access to water while also allowing crops to obtain moisture. It is most effective during dry periods.

Crop Rotation

Rotating crops with ones that kill weeds by choking them out, such as hemp, *Mucuna pruriens*, and other crops, can be a very effective method of weed control. It is a way to avoid the use of herbicides, and to gain the benefits of crop rotation.

Biological Methods

A biological weed control regiment can consist of biological control agents, bioherbicides, use of grazing animals, and protection of natural predators. Post-dispersal, weed seed predators, like ground beetles and small vertebrates, can substantially contribute to the weed regulation by removing weed seeds from the soil surface and thus reduce seed bank size. Several studies provided evidence for the role of invertebrates to the biological control of weeds.

Animal Grazing

Companies using goats to control and eradicate leafy spurge, knapweed, and other toxic weeds have sprouted across the American West.

Chemical Methods

Organic Approaches

Weed control, circa 1930-40s.

A mechanical weed control device: the diagonal weeder.

Organic weed control involves anything other than applying manufactured chemicals. Typically a combination of methods are used to achieve satisfactory control.

Sulfur in some circumstances is accepted within British Soil Association standards.

Herbicides

The above described methods of weed control use no or very limited chemical inputs. They are preferred by organic gardeners or organic farmers.

However weed control can also be achieved by the use of herbicides. Selective herbicides kill certain targets while leaving the desired crop relatively unharmed. Some of these act by interfering with the growth of the weed and are often based on plant hormones. Herbicides are generally classified as follows:

- Contact herbicides destroy only plant tissue that contacts the herbicide. Generally, these are the fastest-acting herbicides. They are ineffective on perennial plants that can re-grow from roots or tubers.
- Systemic herbicides are foliar-applied and move through the plant where they destroy a greater amount of tissue. Glyphosate is currently the most used systemic herbicide.
- Soil-borne herbicides are applied to the soil and are taken up by the roots of the target plant.
- Pre-emergent herbicides are applied to the soil and prevent germination or early growth of weed seeds.

In agriculture large scale and systematic procedures are usually required, often by machines, such as large liquid herbicide 'floater' sprayers, or aerial application.

Bradley Method

Bradley Method of Bush Regeneration, which uses ecological processes to do much of the work. Perennial weeds also propagate by seeding; the airborne seed of the dandelion and the rose-bay willow herb parachute far and wide. Dandelion and dock also put down deep tap roots, which, although they do not spread underground, are able to regrow from any remaining piece left in the ground.

Hybrid

One method of maintaining the effectiveness of individual strategies is to combine them with others that work in complete different ways. Thus seed targeting has been combined with herbicides. In Australia seed management has been effectively combined with trifluralin and clethodim.

Resistance

Resistance occurs when a target adapts to circumvent a particular control strategy. It affects not only weed control,but antibiotics, insect control and other domains.

In agriculture is mostly considered in reference to pesticides, but can defeat other strategies, e.g., when a target species becomes more drought tolerant via selection pressure.

Farming Practices

Herbicide resistance recently became a critical problem as many Australian sheep farmers switched to exclusively growing wheat in their pastures in the 1970s. In wheat fields, introduced varieties of ryegrass, while good for grazing sheep, are intense competitors with wheat. Ryegrasses produce so many seeds that, if left unchecked, they can completely choke a field. Herbicides provided excellent control, while reducing soil disrupting because of less need to plough. Within little more than a decade, ryegrass and other weeds began to develop resistance. Australian farmers evolved again and began diversifying their techniques.

In 1983, patches of ryegrass had become immune to Hoegrass, a family of herbicides that inhibit an enzyme called acetyl coenzyme A carboxylase.

Ryegrass populations were large, and had substantial genetic diversity, because farmers had planted many varieties. Ryegrass is cross-pollinated by wind, so genes shuffle frequently. Farmers sprayed inexpensive Hoegrass year after year, creating selection pressure, but were diluting the herbicide in order to save money, increasing plants survival. Hoegrass was mostly replaced by a group of herbicides that block acetolactate synthase, again helped by poor application practices. Ryegrass evolved a kind of "cross-resistance" that allowed it to rapidly break down a variety of herbicides. Australian farmers lost four classes of herbicides in only a few years. As of 2013 only two herbicide classes, called Photosystem II and long-chain fatty acid inhibitors, had become the last hope.

Strategic Weed Management

Strategic weed management is a process of managing weeds at a district, regional or national scale. In Australia the first published weed management strategies were developed in Tasmania, New South Wales and South Australian 1999, followed by the National Weeds Strategy in 1999.

COVER CROP

A cover crop is planted to manage soil erosion, soil fertility, soil quality, water, weeds, pests, diseases, biodiversity and wildlife in an *agroecosystem*—an ecological system managed and shaped by humans. Cover crops may be an off-season crop planted after harvesting the cash crop. The cover crop may grow over winter.

Soil Erosion

Although cover crops can perform multiple functions in an agroecosystem simultaneously, they are often grown for the sole purpose of preventing soil erosion. Soil erosion is a process that can irreparably reduce the productive capacity of an agroecosystem. Dense cover crop stands physically slow down the velocity of rainfall before it contacts the soil surface, preventing soil splashing and erosive surface runoff. Additionally, vast cover crop root networks help anchor the soil in place and increase soil porosity, creating suitable habitat networks for soil macrofauna. It keeps the enrichment of the soil good for the next few years.

Soil Fertility Management

One of the primary uses of cover crops is to increase soil fertility. These types of cover crops are referred to as "green manure." They are used to manage a range of soil macronutrients and micronutrients. Of the various nutrients, the impact that cover crops have on nitrogen management has received the most attention from researchers and farmers, because nitrogen is often the most limiting nutrient in crop production.

Often, green manure crops are grown for a specific period, and then plowed under before reaching full maturity in order to improve soil fertility and quality. Also the stalks left block the soil from being eroded.

Green manure crops are commonly leguminous, meaning they are part of the pea family, Fabaceae. This family is unique in that all of the species in it set pods, such as bean, lentil, lupins and alfalfa. Leguminous cover crops are typically high in nitrogen and can often provide the required quantity of nitrogen for crop production. In conventional farming, this nitrogen is typically applied in chemical fertilizer form. This quality of cover crops is called fertilizer replacement value.

Another quality unique to leguminous cover crops is that they form symbiotic relationships with the rhizobial bacteria that reside in legume root nodules. Lupins is nodulated by the soil microorganism *Bradyrhizobium* sp. (Lupinus). Bradyrhizobia are encountered as microsymbionts in other leguminous crops (*Argyrolobium, Lotus, Ornithopus, Acacia, Lupinus*) of Mediterranean origin. These bacteria convert biologically unavailable atmospheric nitrogen gas (N_2) to biologically available ammonium (NH^+_4) through the process of biological nitrogen fixation.

Prior to the advent of the Haber-Bosch process, an energy-intensive method developed to carry out industrial nitrogen fixation and create chemical nitrogen fertilizer, most nitrogen introduced to ecosystems arose through biological nitrogen fixation. Some scientists believe that widespread biological nitrogen fixation, achieved mainly through the use of cover crops, is the only alternative to industrial nitrogen fixation in the effort to maintain or increase future food production levels. Industrial nitrogen fixation has been criticized as an unsustainable source of nitrogen for food production due to its reliance on fossil fuel energy and the environmental impacts associated with chemical nitrogen fertilizer use in agriculture. Such widespread environmental impacts include nitrogen fertilizer losses into waterways, which can lead to eutrophication (nutrient loading) and ensuing hypoxia (oxygen depletion) of large bodies of water.

An example of this lies in the Mississippi Valley Basin, where years of fertilizer nitrogen loading into the watershed from agricultural production have resulted in a hypoxic "dead zone" off the Gulf of Mexico the size of New Jersey. The ecological complexity of marine life in this zone has been diminishing as a consequence.

As well as bringing nitrogen into agroecosystems through biological nitrogen fixation, types of cover crops known as "catch crops" are used to retain and recycle soil nitrogen already present. The catch crops take up surplus nitrogen remaining from fertilization of the previous crop, preventing it from being lost through leaching, or gaseous denitrification or volatilization.

Catch crops are typically fast-growing annual cereal species adapted to scavenge available nitrogen efficiently from the soil. The nitrogen tied up in catch crop biomass is released back into the soil once the catch crop is incorporated as a green manure or otherwise begins to decompose.

An example of green manure use comes from Nigeria, where the cover crop *Mucuna pruriens* (velvet bean) has been found to increase the availability of phosphorus in soil after a farmer applies rock phosphate.

Soil Quality Management

Cover crops can also improve soil quality by increasing soil organic matter levels through the input of cover crop biomass over time. Increased soil organic matter enhances soil structure, as well as the water and nutrient holding and buffering capacity of soil. It can also lead to increased soil carbon sequestration, which has been promoted as a strategy to help offset the rise in atmospheric carbon dioxide levels.

Soil quality is managed to produce optimum circumstances for crops to flourish. The principal factors of soil quality are soil salination, pH, microorganism balance and the prevention of soil contamination.

Water Management

By reducing soil erosion, cover crops often also reduce both the rate and quantity of water that drains off the field, which would normally pose environmental risks to waterways and ecosystems downstream. Cover crop biomass acts as a physical barrier between rainfall and the soil surface, allowing raindrops to steadily trickle down through the soil profile. Also, as stated above, cover crop root growth results in the formation of soil pores, which in addition to enhancing soil macrofauna habitat provides pathways for water to filter through the soil profile rather than draining off the field as surface flow. With increased water infiltration, the potential for soil water storage and the recharging of aquifers can be improved.

Just before cover crops are killed (by such practices including mowing, tilling, discing, rolling, or herbicide application) they contain a large amount of moisture. When the cover crop is incorporated into the soil, or left on the soil surface, it often increases soil moisture. In agroecosystems where water for crop production is in short supply, cover crops can be used as a mulch to conserve water by shading and cooling the soil surface. This reduces evaporation of soil moisture. In other situations farmers try to dry the soil out as quickly as possible going into the planting season. Here prolonged soil moisture conservation can be problematic.

While cover crops can help to conserve water, in temperate regions (particularly in years with below average precipitation) they can draw down soil water supply in the spring, particularly if climatic growing conditions are good. In these cases, just before crop planting, farmers often face a tradeoff between the benefits of increased cover crop growth and the drawbacks of reduced soil moisture for cash crop production that season. C/N ratio is balanced with this application.

Weed Management

Cover crop in South Dakota.

Thick cover crop stands often compete well with weeds during the cover crop growth period, and can prevent most germinated weed seeds from completing their life cycle and reproducing. If the cover crop is flattened down on the soil surface rather than incorporated into the soil as a green manure after its growth is terminated, it can form a nearly impenetrable mat. This drastically reduces light transmittance to weed seeds, which in many cases reduces weed seed germination rates. Furthermore, even when weed seeds germinate, they often run out of stored energy for growth before building the necessary structural capacity to break through the cover crop mulch layer. This is often termed the *cover crop smother effect*.

Some cover crops suppress weeds both during growth and after death. During growth these cover crops compete vigorously with weeds for available space, light, and nutrients, and after death they smother the next flush of weeds by forming a mulch layer on the soil surface. For example, researchers found that when using *Melilotus officinalis* (yellow sweetclover) as a cover crop in an improved fallow system (where a fallow period is intentionally improved by any number of different management practices, including the planting of cover crops), weed biomass only constituted between 1-12% of total standing biomass at the end of the cover crop growing season. Furthermore, after cover crop termination, the yellow sweetclover residues suppressed weeds to levels 75–97% lower than in fallow (no yellow sweetclover) systems.

In addition to competition-based or physical weed suppression, certain cover crops are known to suppress weeds through allelopathy. This occurs when certain biochemical cover crop compounds are degraded that happen to be toxic to, or inhibit seed germination of, other plant species. Some well known examples of allelopathic cover crops are *Secale cereale* (rye), *Vicia villosa* (hairy vetch), *Trifolium pratense* (red clover), *Sorghum bicolor* (sorghum-sudangrass), and species in the family Brassicaceae, particularly mustards. In one study, rye cover crop residues were found to have provided between 80% and 95% control of early season broadleaf weeds when used as a mulch during the production of different cash crops such as soybean, tobacco, corn, and sunflower.

In a recent study released by the Agricultural Research Service (ARS) scientists examined how rye seeding rates and planting patterns affected cover crop production. The results show that planting more pounds per acre of rye increased the cover crop's production as well as decreased the amount of weeds. The same was true when scientists tested seeding rates on legumes and oats; a higher density of seeds planted per acre decreased the amount of weeds and increased the yield of legume and oat production. The planting patterns, which consisted of either traditional rows or grid patterns, did not seem to make a significant impact on the cover crop's production or on the weed production in either cover crop. The ARS scientists concluded that increased seeding rates could be an effective method of weed control.

Disease Management

In the same way that allelopathic properties of cover crops can suppress weeds, they can also break disease cycles and reduce populations of bacterial and fungal diseases, and parasitic nematodes. Species in the family Brassicaceae, such as mustards, have been widely shown to suppress fungal disease populations through the release of naturally occurring toxic chemicals during the degradation of glucosinolade compounds in their plant cell tissues.

Pest Management

Some cover crops are used as so-called "trap crops", to attract pests away from the crop of value and toward what the pest sees as a more favorable habitat. Trap crop areas can be established within crops, within farms, or within landscapes. In many cases the trap crop is grown during the same season as the food crop being produced. The limited area occupied by these trap crops can be treated with a pesticide once pests are drawn to the trap in large enough numbers to reduce the pest populations. In some organic systems, farmers drive over the trap crop with a large vacuum-based implement to physically pull the pests off the plants and out of the field. This system has been recommended for use to help control the lygus bugs in organic strawberry production. Another example of trap crops are nematode resistance white mustard (*Sinapis alba*) and radish (*Raphanus sativus*). They can be grown after a main (cereal) crop and trap nematodes, for example the beet cyst nematode and Columbian root knot nematode. When grown, nematodes hatch and are attracted to the roots. After entering the roots they cannot reproduce in the root due to a hypersensitive resistance reaction of the plant. Hence the nematode population is greatly reduced, by 70-99%, depending on species and cultivation time.

Other cover crops are used to attract natural predators of pests by providing elements of their habitat. This is a form of biological control known as habitat augmentation, but achieved with the use of cover crops. Findings on the relationship between cover crop presence and predator/pest population dynamics have been mixed, pointing toward the need for detailed information on specific cover crop types and management practices to best complement a given integrated pest management strategy. For example, the predator mite *Euseius tularensis* (Congdon) is known to help control the pest citrus thrips in Central California citrus orchards. Researchers found that the planting of several different leguminous cover crops (such as bell bean, woollypod vetch, New Zealand white clover, and Austrian winter pea) provided sufficient pollen as a feeding source to cause a seasonal increase in *E. tularensis* populations, which with good timing could potentially introduce enough predatory pressure to reduce pest populations of citrus thrips.

Diversity and Wildlife

Although cover crops are normally used to serve one of the above discussed purposes, they often simultaneously improve farm habitat for wildlife. The use of cover crops adds at least one more dimension of plant diversity to a cash crop rotation. Since the cover crop is typically not a crop of value, its management is usually less intensive, providing a window of "soft" human influence on the farm. This relatively "hands-off" management, combined with the increased on-farm heterogeneity created by the establishment of cover crops, increases the likelihood that a more complex trophic structure will develop to support a higher level of wildlife diversity.

In one study, researchers compared arthropod and songbird species composition and field use between conventionally and cover cropped cotton fields in the Southern United States. The cover cropped cotton fields were planted to clover, which was left to grow in between cotton rows throughout the early cotton growing season (stripcover cropping). During the migration and breeding season, they found that songbird densities were 7–20 times higher in the cotton fields with integrated clover cover crop than in the conventional cotton fields. Arthropod abundance and biomass was also higher in the clover cover cropped fields throughout much of the songbird breeding season, which was attributed to an increased supply of flower nectar from the clover. The clover cover crop enhanced songbird habitat by providing cover and nesting sites, and an increased food source from higher arthropod populations.

BIOLOGICAL PEST CONTROL

Biological control, biocontrol, or biological pest control is a method of suppressing or controlling the population of undesirable insects, other animals, or plants by the introduction, encouragement, or artificial increase of their natural enemies to economically non–important levels. It is an important component of integrated pest management (IPM) programs.

Predatory Polistes wasp looking for bollworms or other caterpillars on a cotton plant.

The biological control of pests and weeds relies on predation, parasitism, herbivory, or other natural mechanisms. Therefore, it is the active manipulation of natural phenomena in serving human purpose, working harmoniously with nature. A successful story of biological control of pests refer to the human beings' capability to depict natural processes for their use and can be the most harmless, non–polluting, and self–perpetuating control method.

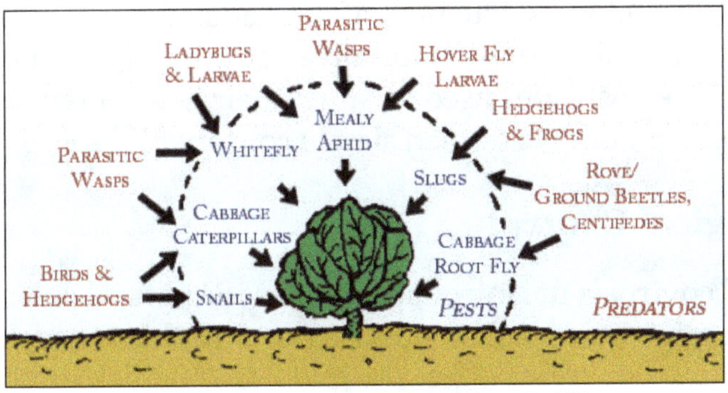

Diagram illustrating the natural enemies of cabbage pests.

In biological control, the reduction of pest populations is achieved by actively using natural enemies.

Natural enemies of the pests, also known as biological control agents, include predatory and parasitoidal insects, predatory vertebrates, nematode parasites, protozoan parasites, and fungal, bacterial, as well as viral pathogens. Biological control agents of plant diseases are most often referred to as antagonists. Biological control agents of weeds include herbivores and plant pathogens. Predators, such as lady beetles and lacewings, are mainly free–living species that consume a large number of prey during their lifetime. Parasitoids are species whose immature stage develops on or within a single insect host, ultimately killing the host. Most have a very narrow host range. Many species of wasps and some flies are parasitoids. Pathogens are disease–causing organisms including bacteria, fungi, and viruses. They kill or debilitate their host and are relatively specific to certain pest or weed groups.

Strategies of Biological Control Methods

There are three basic types of biological control strategies; conservation biocontrol, classical biological control, and augmentative biological control (biopesticides).

Conservation Biocontrol

The conservation of existing natural enemies is probably the most important and readily available biological control practice available to homeowners and gardeners. Natural enemies occur in all areas, from the backyard garden to the commercial field. They are adapted to the local environment and to the target pest, and their conservation is generally simple and cost–effective. For example, snakes consume a lot or rodent and insect pests that can be damaging to agricultural crops or spread disease. Dragonflies are important consumers of mosquitoes.

Eggs, larvae, and pupae of Helicoverpa moths, the main insect pests of cotton, are all attacked by many beneficial insects and research can be conducted in identifying critical habitats, resources needed to maintain them, and ways of encouraging their activity. Lacewings, lady beetles, hover fly larvae, and parasitized aphid mummies are almost always present in aphid colonies. Fungus–infected adult flies are often common following periods of high humidity. These naturally occurring biological controls are often susceptible to the same pesticides used to target their hosts. Preventing the accidental eradication of natural enemies is termed simple conservation.

Classical Biological Control

Classical biological control is the introduction of exotic natural enemies to a new locale where they did not originate or do not occur naturally. This is usually done by government authorities.

In many instances, the complex of natural enemies associated with an insect pest may be inadequate. This is especially evident when an insect pest is accidentally introduced into a new geographic area without its associated natural enemies. These introduced pests are referred to as exotic pests and comprise about 40 percent of the insect pests in the United States. Examples of introduced vegetable pests include the European corn borer, one of the most destructive insects in North America.

To obtain the needed natural enemies, scientists have utilized classical biological control. This is the practice of importing, and releasing for establishment, natural enemies to control an introduced (exotic) pest, although it is also practiced against native insect pests. The first step in the process is to determine the origin of the introduced pest and then collect appropriate natural enemies associated with the pest or closely related species. The natural enemy is then passed through a rigorous quarantine process, to ensure that no unwanted organisms (such as hyperparasitoids or parasites of the parasite) are introduced, then they are mass produced, and released. Follow–up studies are conducted to determine if the natural enemy becomes successfully established at the site of release, and to assess the long–term benefit of its presence.

There are many examples of successful classical biological control programs. One of the earliest successes was with the cottony cushion scale (Icerya purchasi), a pest that was devastating the California citrus industry in the late 1800s. A predatory insect, the Australian lady beetle or vedalia beetle (Rodolia cardinalis), and a parasitoid fly were introduced from Australia. Within a few years, the cottony cushion scale was completely controlled by these introduced natural enemies. Damage from the alfalfa weevil, a serious introduced pest of forage, was substantially reduced by the introduction of several natural enemies like imported ichnemonid parasitoid Bathyplectes curculionis. About twenty years after their introduction, the alfalfa area treated for alfalfa weevil in the northeastern United States was reduced by 75 percent. A small wasp, Trichogramma ostriniae, introduced from China to help control the European corn borer (Pyrausta nubilalis), is a recent example of a long history of classical biological control efforts for this major pest. Many classical biological control programs for insect pests and weeds are under way across the United States and Canada.

Classical biological control is long lasting and inexpensive. Other than the initial costs of collection, importation, and rearing, little expense is incurred. When a natural enemy is successfully established it rarely requires additional input and it continues to kill the pest with no direct help from humans and at no cost. Unfortunately, classical biological control does not always work. It is usually most effective against exotic pests and less so against native insect pests. The reasons for failure are often not known, but may include the release of too few individuals, poor adaptation of the natural enemy to environmental conditions at the release location, and lack of synchrony between the life cycle of the natural enemy and host pest.

Augmentative Biological Control

This third strategy of biological control method involves the supplemental release of natural enemies. Relatively few natural enemies may be released at a critical time of the season (inoculative release) or literally millions may be released (inundative release). Additionally, the cropping system may be modified to favor or augment the natural enemies. This latter practice is frequently referred to as habitat manipulation.

An example of inoculative release occurs in greenhouse production of several crops. Periodic releases of the parasitoid, Encarsia formosa, are used to control greenhouse whitefly, and the predaceous mite, Phytoseilus persimilis, is used for control of the two–spotted spider mite. The wasp Encarsia formosa lays its eggs in young whitefly "scales," turning them black as the parasite larvae pupates. Ideally it is introduced as soon as possible after the first adult whitefly are seen. It is most effective when dealing with low level infestations, giving protection over a long period of time. The predatory mite, Phytoseilus persimilis, is slightly larger than its prey and has an orange body. It develops from egg to adult twice as fast as the red spider mite and once established quickly overcomes infestation.

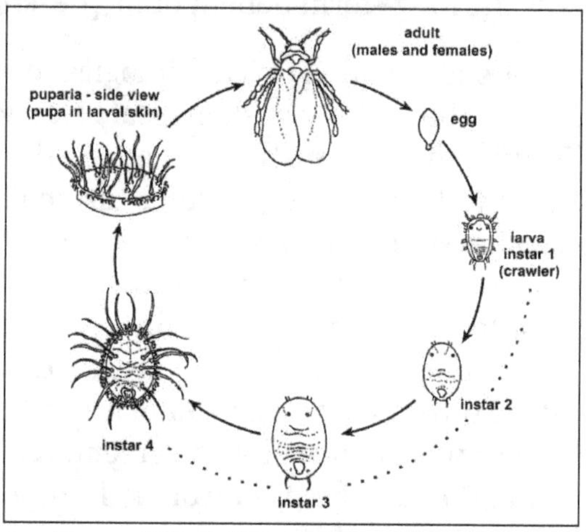

Diagram illustrating the life cycles of Greenhouse whitefly
and its parasitoid wasp Encarsia Formosa.

Lady beetles, lacewings, or parasitoids such as Trichogramma are frequently released in large numbers (inundative release) and are often known as biopesticides. Recommended release rates for Trichogramma in vegetable or field crops range from 5,000 to 200,000 per acre per week depending on level of pest infestation. Similarly, entomoparasitic nematodes are released at rates of millions and even billions per acre for control of certain soil-dwelling insect pests. Entomopathogenic fungus Metarhizium anisopliae var. acridum, which is specific to species of short–horned grasshoppers (Acridoidea and Pyrgomorphoidea) widely distributed in Africa, has been developed as inundative biological control agent.

Habitat or environmental manipulation is another form of augmentation. This tactic involves altering the cropping system to augment or enhance the effectiveness of a natural enemy. Many adult parasitoids and predators benefit from sources of nectar and the protection provided by refuges such as hedgerows, cover crops, and weedy borders. Mixed plantings and the provision of flowering borders can increase the diversity of habitats and provide shelter and alternative food sources. They are easily incorporated into home gardens and even small-scale commercial plantings, but are more difficult to accommodate in large–scale crop production. There may also be some conflict with pest control for the large producer because of the difficulty of targeting the pest species and the use of refuges by the pest insects as well as natural enemies.

Examples of habitat manipulation include growing flowering plants (pollen and nectar sources) near crops to attract and maintain populations of natural enemies. For example, hover fly adults can be attracted to umbelliferous plants in bloom.

Biological control experts in California have demonstrated that planting prune trees in grape vineyards provides an improved overwintering habitat or refuge for a key grape pest parasitoid. The prune trees harbor an alternate host for the parasitoid, which could previously overwinter only at great distances from most vineyards. Caution should be used with this tactic because some plants attractive to natural enemies may also be hosts for certain plant diseases, especially plant viruses that could be vectored by insect pests to the crop. Although the tactic appears to hold much promise, only a few examples have been adequately researched and developed.

Different Types of Biological Control Agents

Predators

Ladybird larva eating wooly apple aphids.

Lacewings are available from biocontrol dealers.

Ladybugs, and in particular their larvae which are active between May and July in the northern hemisphere, are voracious predators of aphids such as greenfly and blackfly, and will also consume mites, scale insects, and small caterpillars. The ladybug is a very familiar beetle with various colored markings, while its larvae are initially small and spidery, growing up to 17 millimeters (mm) long. The larvae have a tapering segmented gray/black body with orange/yellow markings nettles in the garden and by leaving hollow stems and some plant debris over–winter so that they can hibernate over winter.

Hoverflies, resembling slightly darker bees or wasps, have characteristic hovering, darting flight patterns. There are over 100 species of hoverfly, whose larvae principally feed upon greenfly, one larva devouring up to 50 a day, or 1000 in its lifetime. They also eat fruit tree spider mites and small caterpillars. Adults feed on nectar and pollen, which they require for egg production. Eggs are minute (1 mm), pale yellow-white, and laid singly near greenfly colonies. Larvae are 8–17 mm long, disguised to resemble bird droppings; they are legless and have no distinct head. Therefore, they are semi–transparent with a range of colors from green, white, brown, and black. Hoverflies can be encouraged by growing attractant flowers such as the poached eggplant (Limnanthes douglasii), marigolds, or phacelia throughout the growing season.

Dragonflies are important predators of mosquitoes, both in the water, where the dragonfly naiads eat mosquito larvae, and in the air, where adult dragonflies capture and eat adult mosquitoes. Community–wide mosquito control programs that spray adult mosquitoes also kill dragonflies, thus removing an important biocontrol agent, and can actually increase mosquito populations in the long term.

Other useful garden predators include lacewings, pirate bugs, rove and ground beetles, aphid midge, centipedes, as well as larger fauna such as frogs, toads, lizards, hedgehogs, slow–worms, and birds. Cats and rat terriers kill field mice, rats, june bugs, and birds. Dogs chase away many types of pest animals. Dachshunds are bred specifically to fit inside tunnels underground to kill badgers.

Parasitoidal Insects

Most insect parasitoids are wasps or flies. For example, the parasitoid Gonatocerus ashmeadi (Hymenoptera: Mymaridae) has been introduced to control the glassy-winged sharpshooter Homalodisca vitripennis (Hemipterae: Cicadellidae) in French Polynesia and has successfully controlled about 95 percent of the pest density. Parasitiods comprise a diverse range of insects that lay their eggs on or in the body of an insect host, which is then used as a food for developing larvae. Parasitic wasps take much longer than predators to consume their victims, for if the larvae were to eat too fast they would run out of food before they became adults. Such parasites are very useful in the organic garden, for they are very efficient hunters, always at work searching for pest invaders. As adults, they require high–energy fuel as they fly from place to place, and feed upon nectar, pollen and sap, therefore planting plenty of flowering plants, particularly buckwheat, umbellifers, and composites will encourage their presence.

Four of the most important groups are:

- Ichneumonid wasps: (5–10 mm) Prey mainly on caterpillars of butterflies and moths.
- Braconid wasps: Tiny wasps (up to 5 mm) attack caterpillars and a wide range of other insects including greenfly. It is a common parasite of the cabbage white caterpillar, seen as clusters of sulphur yellow cocoons bursting from collapsed caterpillar skin.

- Chalcid wasps: Among the smallest of insects (<3 mm). It parasitizes eggs/larvae of greenfly, whitefly, cabbage caterpillars, scale insects, and strawberry tortrix moth.
- Tachinid flies: Parasitize a wide range of insects including caterpillars, adult and larval beetles, true bugs, and others.

Parasitic Nematodes

Nine families of nematodes (Allantone-matidae, Diplogasteridae, Heterorhabditidae, Mermithidae, Neotylenchidae, Rhabditidae, Sphaerulariidae, Steinernematidae, and Tetradonematidae) include species that attack insects and kill or sterilize them, or alter their development. In addition to insects, nematodes can parasitize spiders, leeches, annelids, crustaceans and mollusks. An excellent example of a situation in which a nematode may replace chemicals for control of an insect is the black vine weevil, Otiorhynchus sulcatus, in cranberries. Uses of chemical insecticides on cranberry either are restricted or have not provided adequate control of black vine weevil larvae. Heterorhabditis bacteriophora NC strain was applied, and it provided more than 70 percent control soon after treatment and was still providing that same level of control a year later.

Many nematode–based products are currently available. They are formulated from various species of Steinernema and Heterorhabditis. Some of the products found in various countries are ORTHO Bio–Safe, BioVector, Sanoplant, Boden-Ntitzlinge, Helix, Otinem, Nemasys, and so forth. A fairly recent development in the control of slugs is the introduction of "Nemaslug," a microscopic nematode (Phasmarhabditis hermaphrodita) that will seek out and parasitize slugs, reproducing inside them and killing them. The nematode is applied by watering onto moist soil, and gives protection for up to six weeks in optimum conditions, though is mainly effective with small and young slugs under the soil surface.

Plants to Regulate Insect Pests

Choosing a diverse range of plants for the garden can help to regulate pests in a variety of ways, including:

- Masking the crop plants from pests, depending on the proximity of the companion or intercrop.
- Producing olfactory inhibitors, odors that confuse and deter pests.
- Acting as trap plants by providing an alluring food that entices pests away from crops.
- Serving as nursery plants, providing breeding grounds for beneficial insects.
- Providing an alternative habitat, usually in a form of a shelterbelt, hedgerow, or beetle bank, where beneficial insects can live and reproduce. Nectar–rich plants that bloom for long periods are especially good, as many beneficials are nectivorous

during the adult stage, but parasitic or predatory as larvae. A good example of this is the soldier beetle, which is frequently found on flowers as an adult, but whose larvae eat aphids, caterpillars, grasshopper eggs, and other beetles.

Pathogens to be used as Biopesticides

Various bacterial species are widely used in controlling the pests as well as weeds. The best–known bacterial biological control which can be introduced in order to control butterfly caterpillars is Bacillus thuringiensis, popularly called Bt. This is available in sachets of dried spores, which are mixed with water and sprayed onto vulnerable plants such as brassicas and fruit trees. After ingestion of the bacterial preparation, the endotoxin liberated and activated in the midgut will kill the caterpillars, but leave other insects unharmed. There are strains of Bt that are effective against other insect larvae. Bt. israelensis is effective against mosquito larvae and some midges.

Viruses most frequently considered for the control of insects (usually sawflies and Lepidoptera) are the occluded viruses, namely NPV, cytoplasmic polyhedrosis (CPV), granulosis (GV), and entomopox viruses (EPN). They do not infect vertebrates, non–arthropod invertebrates, microorganisms, and plants. The commercial use of virus insecticides has been limited by their high specificity and slow action.

Fungi are pathogenic agents to various organisms including the pests and the weeds. This feature is intensively used in biocontrol. The entomopathogenic fungi, like Metarhizium anisopliae, Beauveria bassiana, and so forth cause death to the host by the secretion of toxins. A biological control being developed for use in the treatment of plant disease is the fungus Trichoderma viride. This has been used against Dutch Elm disease, and to treat the spread of fungal and bacterial growth on tree wounds. It may also have potential as a means of combating silver leaf disease.

Significance of Biological Control

Biological control proves to be very successful economically, and even when the method has been less successful, it still produces a benefit–to–cost ratio of 11:1. The benefit–to–cost ratios for several successful biological controls have been found to range from 1:1 to 250:1. Further, net economic advantage for biological control without scouting vs. conventional insecticide control ranged from \$ 7.43 to \$ 0.12 per hectare in some places. It means that even if the yield produce under biological control be below that for insecticidal control by as much as 29.3 kilos per hectare, the biological control would not lose its economic advantage.

Biological control agents are non–polluting and thus environmentally safe and acceptable. Usually they are species specific to targeted pest and weeds. The biological control discourages the use of environmentally and ecologically unsuitable chemicals, so it always leads to the establishment of natural balance. The problems of increased resistance in the pest will not arise, as both biological control agents and the pests are in complex race of evolutionary dynamism. Because of chemical resistance developed by the Colorado potato beetle (CPB), its control has been achieved by the use of bugs and beetles (Hein).

Trap Crop

A trap crop is a plant that attracts agricultural pests, usually insects, away from nearby crops. This form of companion planting can save the main crop from decimation by pests without the use of pesticides. While many trap crops have successfully diverted pests off of focal crops in small scale greenhouse, garden and field experiments, only a small portion of these plants have been shown to reduce pest damage at larger commercial scales. A common explanation for reported trap cropping failures, is that attractive trap plants only protect nearby plants if the insects do not move back into the main crop. In a review of 100 trap cropping examples in 2006, only 10 trap crops were classified as successful at a commercial scale, and in all successful cases, trap cropping was supplemented with management practices that specifically limited insect dispersal from the trap crop back into the main crop.

Usage

Trap crops, when used on an industrial scale, are generally planted at a key time in the pest's life-cycle, and then destroyed before that life-cycle finishes and the pest might have transferred from the trap plants to the main crop.

Examples of trap crops include:

- Alfalfa planted in strips among cotton, to draw away lygus bugs, while castor beans surround the field, or tobacco is planted in strips among it, to protect from the budworm Heliothis.
- Rose enthusiasts often plant Pelargonium geraniums among their rosebushes because Japanese beetles are drawn to the geraniums, which are toxic to them.
- Chervil is used by gardeners to protect vegetable plants from slugs.
- Rye, sesbania, and sicklepod are used to protect soybeans from corn seeding maggots, stink bugs, and velvet green caterpillars, respectively.
- Mustard and Alfalfa planted near strawberries to attract lygus bugs, a method pioneered by Jim Cochran.

Trap crops can be planted around the circumference of the field to be protected, which is assumed to act as a barrier for entry by pests, or they can be interspersed among the focul crop, for example being planted every ninth row. Planting crops in rows helps facilitate supplemental management practices that prevent insect pest dispersal back into the main field, such as driving a vehicle above the trap crop which then removes insect pests by vacuuming them off of the trap crop row or targeted insecticides, which are only deployed on the trap crop. Even if pesticides are used to control insects on the trap crop, total pesticides are greatly reduced in this scenario over conventional agricultural pesticide applications because they are only deployed on a small portion of the farm (the trap crop). Other strategies that prevent dispersal of insect pests back into the main crop include cutting the trap plants, applying predators to the trap plant that eat the pest, and planting a high ratio of trap plants to other plants.

Operation

Recent studies on host-plant finding have shown that flying pests are far less successful if their host-plants are surrounded by any other plant, or even "decoy-plants" made of green plastic, cardboard or any other green material. The host-plant finding process occurs in three phases.

The first phase is stimulation by odours characteristic to the host-plant. This induces the insect to try to land on the plant it seeks. But insects avoid landing on brown (bare) soil. So if only the host-plant is present, the insects will quasi-systematically find it by landing on the only green thing around. This is called an "appropriate landing". When it does an "inappropriate landing", it flies off to any other nearby patch of green. It eventually leaves the area if there are too many "inappropriate" landings.

The second phase of host-plant finding is for the insect to make short flights from leaf to leaf to assess the plant's overall suitability. The number of leaf-to-leaf flights varies according to the insect species and to the host-plant stimulus received from each leaf. But the insect must accumulate sufficient stimuli from the host-plant to lay eggs; so it must make a certain number of consecutive "appropriate" landings. Hence if it makes an "inappropriate landing", the assessment of that plant is negative and the insect must start the process anew.

Thus it was shown that clover used as a ground cover had the same disruptive effect on eight pest species from four insect orders. An experiment showed that 36% of cabbage root flies laid eggs beside cabbages growing in bare soil (which resulted in no crop), compared with only 7% beside cabbages growing in clover (which allowed a good crop). Also that simple decoys made of green card disrupted appropriate landings just as well as did the live ground cover.

Beetle Bank

Beetle banks mimic the dense grass bases of traditional hedgerows and provide shelter for a great deal of valuable and beneficial farm wildlife.

Beetle banks are grass mounds that cut across the middle of large arable fields, providing permanent vegetation cover for beetles and other beneficial insects.

Typically beetle banks are 2m wide and run the length of arable fields, but can start from 20m out from the field edge to allow fields to still be managed as a single unit.

Beetle banks can also be strategically placed across slopes of fields prone to surface run-off to improve infiltration and reduce the risk of soil erosion.

Beetle banks are generally appropriate for fields larger than 20ha. Fields of 30ha and above may warrant more than one beetle bank.

Beetle banks provide essential over-winter habitat for beneficial predatory insects and spiders which will move into the crop in the spring and feed on crop pests. Beetle banks can be a valuable source of these pest predators, particularly in the centre of large fields as typically beetles and spiders will not travel more than 250m from field margins. The habitat provided by beetle banks can be even more important in landscapes where there are fewer hedgerows and the associated base of dense grasses which would ordinarily offer shelter for beneficial insects.

Beetle banks can also provide habitat for ground-nesting birds, small mammals and amphibians and reptiles. They are preferred by species that nest in open farmland away from field boundaries, such as corn buntings and skylarks, and mammals such as harvest mice. Tussocky banks may also provide nesting and over-winter habitat for bumblebees.

August and September are the best months to establish the grass sward that forms a beetle bank.

Beetle banks are created by ploughing furrows that point towards each other during cultivation to create a raised bank of about 0.4m along the plough line. The headland at each end of the beetle bank can remain cropped. Leaving a gap of no more than 25m between the end of the beetle bank and the field margin will reduce ground predators such as stoats accessing the beetle bank. The bank should be seeded with a native grass seed mix with up to 30% tussock-forming species such as cocksfoot or timothy. The rest of the mix can consist of fescues and bents. The seed should be broadcast immediately after cultivation at a rate of 70kg/ha for good establishment. Although more costly, mixes can include tall, tussock-tolerant wildflowers which can encourage other pest predators such as hoverflies and parasitic wasps.

In the first summer, 3 cuts may be necessary when the sward reaches 10cm in height, to encourage grasses to tiller and to help control invasive annual weeds. Once established, cut the grass strips as little as possible as established tussocks are key overwintering features for many of the species that will utilise the beetle bank. Cutting on a 3 year rotation is ideal, and this will help to produce a diverse grass structure which will support a wider range of wildlife.

Cutting should take place during autumn, before insects look to find shelter over winter.

To improve the degree of natural pest control in large arable fields, beetle banks should be located through the middle of fields larger than 20ha, between tramlines so that the crop can still be sprayed on either side with minimal disruption to spraying operations. To intercept surface run-off, beetle banks should run along the contours of the slope so that the raised bank holds up water running down the slope and increase the rate of infiltration into the soil.

Beetle banks are particularly vulnerable to the effects of pesticide drift where they are sprayed on either side. This can reduce their ability to harbour beneficial insects and spiders. Adding grass buffer strips to both or either sides can help to alleviate this. On light soils with a low weed burden, conservation headland management on either side of a beetle bank will enhance its value for insects and birds whenever the adjacent crop is a cereal. Alternatively, more habitat diversity can be created by establishing flower-rich habitats on the adjoining sides to further bolster populations of beneficial insects.

The genus Verticillium is composed of a small group of ascomycete fungi belonging to division Ascomycota. As such, they may also be described as sac fungi.

Today, about ten species of this genus have been clearly identified, some of which are phytopathogenic (plant pathogens). The plant-pathogenic species in the genus have been shown to cause damage to a wide variety of crops and can be very costly to control.

Some of the most important species in this genus include:

- Verticillium dahliae,
- Verticillium albo-atrum,
- Verticillium lecanii.

Some of the characteristics of species in this genus include:

- They have/form an ascus (and may therefore be described as asci fungi),
- They form resting structures.

While some live as saprophytes, a good number depend on plants for part of their life cycle:

- Ovoid/ellipsoid conidiospores,
- Conidiospores are single celled,
- Long phialides,
- The vegetative mycelium has the following characteristics: hyaline, multinucleate and septet.

Verticillium Dahliae

Along with a number of other species (such as V. albo-atrum), V. dahliae is a soil borne pathogen that is classified under class Deuteromycetes, also referred to as Fungi Imperfecti. It affects and causes wilt in a wide variety of host crops including cotton, potato and fruits like strawberry among others.

Apart from barley, studies have shown monocotyledonous plants to be non-hosts, which means that the pathogen rarely affects monocots (plants whose seed in the seed coat contains one embryonic leaf). Because they can be found in many regions across the world (temperate and subtropical zones as well as in some tropic regions) V. dahliae causes significant crop damage globally costing farmers heavy losses.

Life Cycle

The life cycle of V. dahliae can be divided into three major phases that include:

- Dormant phase,
- Parasitic phase,
- Saprotrophic phase.

The dormant stage of V. dahliae refers to the phase or extended period of time in which the pathogen remains dormant. During this period, a round, thick-walled structure known as microsclerotia made of melanin allows the fungus (plural fungi) to survive in soil during unfavorable conditions for a period that may range from one year to as long as two decades. During this period, the microsclerotia remains viable and can germinate at any given time when conditions are favorable.

While the mechanism is yet to be clearly understood, the microsclerotia of V. dahliae has been shown to germinate in the presence of root exudates (part of plant defense around the roots) in the rhizosphere of the host plants. As they germinate, the fungi increasingly infects the tips of the roots (of host plants) with varying colonization of the plant cortex from one plant to another.

In plants where the fungi has successfully colonized the cortex, they grow into the vascular system where the conidia is formed. This is of great benefit for the fungi given that the vascular system allows the conidia to be transported to other vessels in the plant through the sap stream. In the process, the fungi conidia are transported to other parts of the plant.

Studies have also shown the fungi to produce various enzymes (pectinolytic enzymes) that actively cause damage to the parts of the xylem among other sections of the vessels. A glue-like substance is also produced that not only causes the discoloration of the vessels and parenchymal cells of the plant, but also clogs the vessels as they are continuously filled with fungal hyphae.

Such toxins as the lipopolysaccharides are also produced which in addition to the other mechanisms, affect photosynthesis and respiration and ultimately result in wilting. This is the parasitic phase/stage.

The parasitic phase/stage of the life cycle significantly affects the host plant and enhances its deterioration (senescence). This is followed by the saprophytic stage as the plant deteriorates further and dies.

During this phase/stage, the fungi increasingly colonizes various parts of the plant (particularly the roots and shoot) with increased production of the microsclerotia. This stage allows the fungi to continue growing and reproducing as more microsclerotia are produced in the cortex and petioles of the plant.

Once the plant completely dies and degrades, the microsclerotia left in the soil can continue with the life cycle when they are stimulated by new root exudates.

Morphology of Verticillium Dahliae

When viewed in culture media, the following traits of V. dahliae were identified:

- Short and dark hyphae (Some of the microsclerotia had hyphae growing from them).
- Dark mycelium.
- Elongated or globose microsclerotia.

Verticillium dahliae as well as other species like Verticillium longisporum tend to cause significant crop damage that results in significant losses on the part of farmers.

Infection is easily detected through such symptoms as necrosis and tissue death, half-side discoloration of the leaves as well as tissue discoloration. For this reason, many farmers tend to avoid more susceptible plants or control the pathogen through crop rotation.

Soil fumigation that was once used to control the pathogen is no longer allowed in mane country due to the negative impacts of the chemical thus the lack of chemical control for the fungus.

While there is no chemical control for the fungus, studies are being conducted to develop the ideal chemical agent that will control the fungus without causing environmental problems/pollution.

Verticillium Lecanii

Unlike V. dahliae, V. lecanii is an entomopathogenic fungus, which means that it acts as a parasite of various insects.

V. lecanii has a wide variety of insect hosts that range from aphids, whiteflies and leafhoppers (of sub-order Homoptera) to locusts, crickets and grasshoppers that belong to the order Orthoptera. This is in addition to parasitizing various plant pathogens such as soybean cyst nematodes. Because of their ability to parasitize and even kill these insects, V. lecanii has become one of the most important biological agents used for pest and insect control.

Previously known as Verticillium lecanii Zimm, the fungus is today simply referred to as L. lecanii belonging to the Superkingdom Eukaryota under the Kingdom Fungi. It also falls under Phylum Ascomycota and Family Hypocreaceae.

Life Cycle of V. Lecanii

The life cycle of V. lecanii starts with the conidia (spores) landing and adhering to the protective outer layer of the insect (epicuticle).

While on this outer protective layer, the conidia starts germinating and forming elongated germ tubes capable of penetrating the hard layer. This is made possible by the secretion of enzymes that enhance the lysis of the epicuticle making it possible for the germ to easily penetrate the covering and continue growing in the host insect.

Some of the enzymes (extra-cellular enzymes) produced to help in the penetration include:

- Aminopeptidase lipase,
- Proteases,
- Esterase,
- Chitinases.

These enzymes are released on the cuticle and start digesting it to allow the germ tube to penetrate and grow into the insect.

The germ tube then grows into hyphae that continues growing and colonizing the tissues of the insect. As the hyphae grows, it causes the insect to change color, ranging from yellow to red.

Ultimately, the insect dies from the pressure caused by hyphal growth within the insect as well as the enzymes secreted. Nutrition from the tissue of the insect allows the hyphae to continue growing and finally differentiate to conidiophores on the surface of the insect that produce conidia. Once the conidia lands on the epicuticle of another host insect, the life cycle begins.

When viewed under the microscope following culture, the following traits were identified:

- Septate hyaline hyphae,
- The hyphae were either simple or branched,

- Cylindrical hyphae some of which has pointed ends,
- Elongated phialides,
- Oval hyaline.

Pest Control

Whereas V. dahliae is undesired for its effects on a wide range of crops, the appeal of V. Lecanii lies in the mechanism of its life cycle.

As already mentioned, V. Lecanii is an entomopathogenic fungus that grows and reproduces inside an insect and ultimately kills it. This has made it one of the most important fungi used for pest control.

While it may be formulated in different ways by different companies, the formulation allows for the spores to remain viable so that they can grow on and kill host pests (aphids, whiteflies etc). For instance, the wettable powder formulation carries viable spores that are capable of penetrating the cuticle of insects such as aphids and killing them in the farm and greenhouses.

Some of the other pests/insects targeted by this biological agent include:

- Pod flies,
- Mites,
- Flower webber,
- Hoppers (leaf hoppers, grass hoppers etc).

References

- Crop-protection-methods, crops: bionovelus.com, Retrieved 6 January , 2019
- "Chromated Copper Arsenate (CCA): Cyproconazole - An Alternative to CCA". EPA. 2 July 2014. Retrieved 4 June 2015
- Herbicide, science: britannica.com, Retrieved 13 May, 2019
- Cyproconazole, Pesticide Properties database, University of Hertfordshire, U.K., 26 June 2015, accessed 19 September 2015
- "U.S. EPA settles with agribusiness firm for $15,600 for failing to report pesticides produced in India and Switzerland". News Releases from Region 9. US EPA. 20 September 2006. Retrieved 19 September 2015
- Van Emden, H.F.; Peakall, David B. (30 June 1996). Beyond Silent Spring. Springer. ISBN 978-0-412-72800-6
- Kupferschmidt, K. (2013). "A Lethal Dose of RNA". Science. 341 (6147): 732–3. Bibcode:2013Sci...341..732K. Doi:10.1126/science.341.6147.732. PMID 23950525

- Golden Gate Gardening: For more information on bird protection net / anti bird net refer birdproofingsolutions.in The Complete Guide to Year-Round Food Gardening (2002) ISBN 1-57061-136-X, p. 151
- Crop-disease-manual: croppro.com.au, Retrieved 5 August, 2019
- J. C. Van Lenteren (2003). Quality Control and Production of Biological Control Agents: Theory and Testing Procedures. CABI. ISBN 978-0-85199-836-7
- Beetle-banks, farmed-area, how-to-do-it: farmwildlife.info, Retrieved 3 February, 2019
- Van Lenteren, J. C. (2009). "Implementation of biological control". American Journal of Alternative Agriculture. 3 (2–3): 102. Doi:10.1017/S0889189300002265
- "Control methods". Department of Agriculture and Food, Government of Western Australia. Retrieved 11 December 2015
- Biological-pest-control, entry: newworldencyclopedia.org, Retrieved 4 June, 2019
- Alfredo Flores (March 8, 2010). "Geraniums Could Help Control Devastating Japanese Beetle". United States Department of Agriculture Agricultural Research Service. Retrieved 2010-04-04
- Verticillium-fungus: microscopemaster.com, Retrieved 26 August, 2019
- Samuel Fromartz (2005). Organic Inc.: Natural Foods and How They Grew. Harcourt. ISBN 978-0-15-603242-1

Crop Management

CHAPTER 6

Crop management refers to the agricultural practices which are used to improve the development, growth and yield of agricultural crops. Some of the important practices associated with it are water management and crop rotation. All these diverse aspects of crop management have been carefully analyzed in this chapter.

There are various methods of crop management practices, varying according to the types of resources used, climate and various other natural factors that involves the type of management to choose. Crop management involves growing crops on a massive scale, by involving methods like tillage, crop rotation, inter cropping, culling, composting, pest control and weed management, effective resources capture, irrigation techniques and hybridization.

Weeds, Pests and Disease Control

The European Weed Science Society defines weed as 'any plant or vegetation, excluding fungi, interfering with the objectives or requirements of people'. Reduction of crop yield is the major reason for attempts to reduce weed population in arable crops. The first attempts at controlling weeds used manual labour and hand-hoeing which, at present days is a time consuming process. Thus a major development was the development of herbicides.

The worldwide efficacy in controlling weed (68%) was considerably higher than the control of animal pests or diseases (39% and 32%, respectively). The continuous use of herbicide and pesticides led to the appearance of resistant biotypes of weeds and pests, playing major role in contamination of the environment.

In the recent days the integrated weed management (IWM) approaches plays a vital role as they incorporate multiple tactics of prevention, avoidance, monitoring and suppression of weeds, undergirded by the knowledge of agro ecosystem biology. The development of IWM was motivated by a desire to provide farmers with systematic approaches to reduce reliance upon herbicide and retard the selection of herbicide-resistant biotypes.

Worldwide, pre-harvest and post-harvest losses to insects, weeds, and plant pathogens are estimated at 45 percent additional losses can be attributed to vertebrate pests. Selection of resistant varieties, rotation crops, crops for inter plantings and cover crops is important for successful crop production with minimal inputs of chemical pesticides and fertilizers. Studies say that the use of pesticides has not solved the pest problem completely and in the past 50 years, insecticide use has increased tenfold, while crop losses from pest damage have doubled. Thus still there are studies being carried on pest control. The natural methods of control is more preferable for three major reasons like cost, safe to people and safe to environment also has no side effects.

Pests and diseases impact on crop yield and quality, and also reduce resource-use efficiency. Improved crop protection strategies to prevent such damage and loss can increase production and make a substantial contribution to food security.

Irrigation Technique

Irrigation is the method of artificial application of water to the land or soil. It is used to assist in the growing of agricultural crops, maintenance of landscapes, and rejuvenation of disturbed soils in dry areas and during periods of inadequate rainfall. The basic principle of irrigation is to help plants on evapo-transpiration process which plays a vital role in transportation of nutrients from soil to any part of the plant.

More Yield more Profit

There are various types of irrigation techniques varying upon the climate, nature of soil, nature of vegetation. In general, the goal is to supply the entire field uniformly with water, so that each plant has the amount of water it needs, neither too much nor too little. The earliest methods of irrigation were the perennial, basin and the terrace irrigation and the recent day's developments were the sprinkler (gun sprinkler, portable solid-set sprinkler, pivot systems, side-roll sprinkler etc.) and micro irrigation (drip irrigation). The type of irrigation technique plays a major role in the crop yield. But each irrigation systems have its own limitations and thus it is very important to choose the most suitable irrigation technique for particular variety of crop, after studying its various properties on its growth and development. For example, a study made in the western United States corn belt showed variation in the yield differing between the pivot and surface irrigation. Therefore choosing the right way of irrigation method is also an important way of crop management.

Resource Capture

The principle of resource capture is to provide a framework through which the productivity of vegetation can be analysed through the capture and conversion of environmental resources. The most basic resources that a plant would obtain are from the atmosphere and soil. The plants obtain carbon dioxide, light, temperature, and humidity from the atmosphere, whereas obtains water and other mineral nutrients from the soil. So resource capture is classified as the above ground and below ground resource capture.

The crops depend on light for growth and depend on day length and temperature for its development. In the below ground resource capture, key natural resource which provides the most essential forms of nutrients for the plants growth which is associated with the integrated effect of management on most properties of soil that determines the sustainability and crop productivity is the 'soil'. Good soil quality not only produces good crop yield, but also maintains environmental quality and consequently plant, animal and human health. Growing of crops one after another without giving due consideration to nutrient requirement has resulted in decline in soil fertility, especially of nitrogen. Studies say that Zero- or reduced tillage crop production practices, coupled with proper residue management can maintain or improve soil organic matter and has the potential to substantially increase long-term crop production in semi-arid rain fed regions.

CROPPING TECHNIQUES

Plants are important segment of life system. These are not only a major component for construction of biodiversity but also utilized for various purpose among the human beings. Crops are diversified in different ecological areas. As per demand and plant adaption in different zones of natural ecosystem these are showing their presence and are cultivated for certain purposes like for fuel, food, fodder, medicine etc. The rate of crop production is variable based on the several biotic and abiotic factors. Overall an environmental effect on production rate is remarkable for individual crop species cultivated by the peoples for specific targets. Plants are also a major source of food, oil, pulses, cereals etc. Basics of cropping techniques requires some important steps to follow for successful growth and development of plants which further leads to better crop production in certain cropping crops.

Step 1: Selection of the Cropping Field

It is a first stage for cropping of any crop in the field. A better field selection should be done on the basis of better soil quality, water facility, easy to reach etc. A field should be carefully selected as per demand of cropping plants to support rich production.

Step 2: Preparation of the Cropping Field

Secondly, just after selection of the suitable field there is a need to prepare the field well for certain crop varieties. Better ploughing, labeling, bed formation, weeding etc. are important practices which should be properly done following suitable techniques which further utilized as a better platform for different crop species.

Step 3: Selection of Crop Species

As per adaption of the crop in particular climatic zones and their demand these should be identified and selected to grow in prepared fields.

Step 4: Preparation of the Crop Plants

Selected crop varieties need for proper preparation to grow in suitable time of cropping. Crop plants can be propagated either by their mature seeds or by utilizing their vegetative parts/modified plant structures such as bulb, tuber, rhizome, corm etc. All the plant materials found to be suitable to propagate further in cropping field should be prepared well following suitable techniques such as seeds should be treated before cultivation and modified plant parts need to develop new buds which further become capable to development new plants like their parental ones. All plant propagules should be prepared/developed well before of cropping. It is done for their maximum germination/ development of new plant species in the field.

Step 5: Sowing/Cultivation

In prepared fields plant propagules should be sown/cultivated following better techniques for rich plant population development in the field which further leads to rich plant development.

Step 6: Management

Water

Just after of cropping there is an urgent need of light irrigation in cropping field for initiation of embryonic growth, bud initiation etc. Water should be supplied as per need of the crops in the field.

Weed

These are undesirable and unwanted plants in cropping fields which grows naturally. Weeding should be done carefully without damaging of the mother plants.

Nutrients

Each plant requires micro and macro nutrients to support their growth and development as well as for production. So, it should be supplied in the field in balanced manner.

Protection

The crops which are grown in the field should be protected against various biotic and abiotic agents and also from diseases to minimize the crop yields loss.

Step 7: Harvesting

At the endings of life or after maturation of crops are indicating by their morphological changes like colour changes from green to yellow brown. Harvesting should be done in suitable time and techniques without damaging of the plants but some crops like paddy etc. are completely harvested. Better harvested plant products registered better prize in market.

Step 8: Collection

Harvested plant materials should be collected carefully, it should be made sundry to reduce the moisture level and toi prevent microbial infection.

Step 9: Storage

Plants products are dried and stored without damaging grains/plant parts. After utilizing a specific amount of Plant products are to be sold in market in better prize. Finally, it is concluded that above described steps are important channel which not only plays an important role in better plant growth but also significant in crop production.

AGRICULTURE WATER MANAGEMENT

The main challenge confronting water management in agriculture is to improve water use efficiency and its sustainability.

This can be achieved through (i) an increase in crop water productivity (an increased in marketable crop yield per unit of water transpired) through irrigation, (ii) a decrease in water losses through soil evaporation that could otherwise be used by plants for their growth, and (iii) an increase in soil water storage within the plant rooting zone through better soil and water management practices at farm and area-wide (catchment) scales.

Tracking and quantifying water fluxes at different spatial and temporal scales within the plant rooting zone remains a formidable challenge because of the interactions between water sources from rainfall, irrigation and subsurface water on plant uptake, soil evaporation, plant transpiration (water transpired by plants) and runoff or drainage losses from crop-growing areas. The use of isotopic and nuclear techniques to investigate the relative importance of soil and irrigation management factors that influence these interactions will greatly assist in the development of water management packages that involve the consideration of soil nutrient status, type of crops grown, growth stages and the overall agro-ecosystems to minimize not only water but also nutrient losses from the farmlands and enhance water and nutrient use efficiencies in agro-ecosystems under both rainfed and irrigated conditions.

Many nuclear and isotopic techniques are being employed in soil water management studies. The soil moisture neutron probe is ideal in field-scale rooting zone measurement of soil water, providing accurate data on the availability of water for determining crop water use and water use efficiency and for establishing optimal irrigation scheduling under different cropping systems especially under saline conditions.

The use of oxygen-18, hydrogen-2 (deuterium) and other isotopes is an integral part of agricultural water management, allowing the identification of water (and plant nutrients) sources and the tracking of water movement and pathways within agricultural landscapes as influenced by different irrigation technologies, cropping systems and farming practices. It also helps in the understanding of plant water use, quantifying crop transpiration and soil evaporation and allows us to devise strategies to improve crop production, reduce unproductive water losses and prevent land and water degradation.

IRRIGATION

Irrigation is the application of controlled amounts of water to plants at needed intervals. Irrigation helps to grow agricultural crops, maintain landscapes, and revegetate disturbed soils in dry areas and during periods of less than average rainfall. Irrigation also has other uses in crop production, including frost protection, suppressing weed growth in grain fields and preventing soil consolidation. In contrast, agriculture that relies only on direct rainfall is referred to as rain-fed or dry land farming.

Irrigation systems are also used for cooling livestock, dust suppression, disposal of sewage, and in mining. Irrigation is often studied together with drainage, which is the removal of surface and sub-surface water from a given area.

Irrigation canal in Osmaniye, Turkey.

Sprinkler irrigation of blueberries in Plainville, New York, United States.

Irrigation has been a central feature of agriculture for over 5,000 years and is the product of many cultures. Historically, it was the basis for economies and societies across the globe, from Asia to the Southwestern United States.

Irrigation of land in Punjab, India.

In year 2000, the total fertile land was 2,788,000 km^2 (689 million acres) and it was equipped with irrigation infrastructure worldwide. About 68% of this area is in Asia, 17% in the Americas, 9% in Europe, 5% in Africa and 1% in Oceania. The largest contiguous areas of high irrigation density are found:

- In Northern India and Pakistan along the Ganges and Indus rivers,
- In the Hai He, Huang He and Yangtze basins in China,
- Along the Nile river in Egypt and Sudan,
- In the Mississippi-Missouri river basin, the Southern Great Plains, and in parts of California.

Smaller irrigation areas are spread across almost all populated parts of the world.

By 2012, the area of irrigated land had increased to an estimated total of 3,242,917 km^2 (801 million acres), which is nearly the size of India.

Advantages of Irrigation

- Increase in food production: Increase in crop yield leads to an increase in food production, thus developing people as well as society.
- Protection against drought: The provision of adequate watering facilities in any region ensures protection against failure of crops from famines and droughts.
- Revenue generation: When the regular supply of water is assured, the farmers can grow certain superior or high priced crops in place of inferior or low priced crops. As a result, revenue is generated.
- Mixed Cropping: Means growing two or more crops together in the same field. This practice is followed so that if weather conditions are not favourable for one crop, it may be suitable for other crops. But if enough water facilities are made available, the need for mixed cropping is eliminated.

Types of Irrigation

There are several methods of irrigation. They vary in how the water is supplied to the plants. The goal is to apply the water to the plants as uniformly as possible, so that each plant has the amount of water it needs, neither too much nor too little.

Surface Irrigation

Basin flood irrigation of wheat.

Surface irrigation is the oldest form of irrigation and has been in use for thousands of years. In *surface* (*flood*, or *level basin*) irrigation systems, water moves across the surface of an agricultural lands, in order to wet it and infiltrate into the soil. Surface irrigation can be subdivided into furrow, *border strip or basin irrigation*. It is often called *flood irrigation* when the irrigation results in flooding or near flooding of the cultivated land. Historically, this has been the most common method of irrigating agricultural land and is still used in most parts of the world.

Where water levels from the irrigation source permit, the levels are controlled by dikes, usually plugged by soil. This is often seen in terraced rice fields (rice paddies), where the method is used to flood or control the level of water in each distinct field. In some cases, the water is pumped, or lifted by human or animal power to the level of the land. The water application efficiency of surface irrigation is typically lower than other forms of irrigation.

Residential flood irrigation in Phoenix, Arizona.

Surface irrigation is even used to water landscapes in certain areas, for example, in and around Phoenix, Arizona. The irrigated area is surrounded by a berm and the water is delivered according to a schedule set by a local irrigation district.

Micro-Irrigation

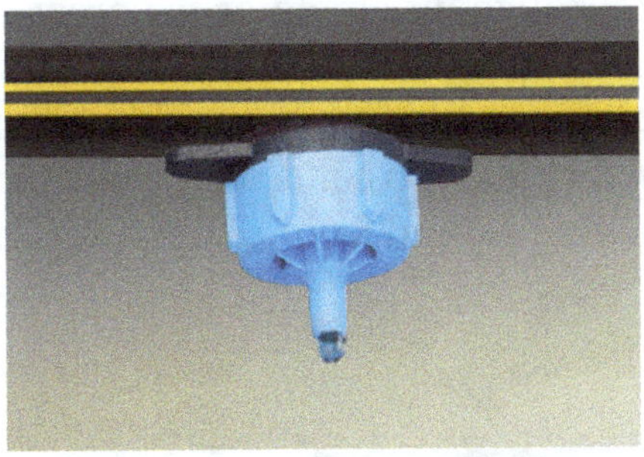

Drip irrigation – a dripper in action.

Micro-irrigation, sometimes called localized irrigation, low volume irrigation, or trickle irrigation is a system where water is distributed under low pressure through a piped network, in a pre-determined pattern, and applied as a small discharge to each plant or adjacent to it. Traditional drip irrigation using individual emitters, subsurface drip irrigation (SDI), micro-spray or micro-sprinkler irrigation, and mini-bubbler irrigation all belong to this category of irrigation methods.

Drip Irrigation

Micro-sprinkler.

Drip (or micro) irrigation, also known as trickle irrigation, functions as its name suggests. In this system water falls drop by drop just at the position of roots. Water is delivered at or near the root zone of plants, drop by drop. This method can be the most water-efficient method of irrigation, if managed properly, evaporation and runoff are minimized. The field water efficiency of drip irrigation is typically in the range of 80 to 90 percent when managed correctly.

In modern agriculture, drip irrigation is often combined with plastic mulch, further reducing evaporation, and is also the means of delivery of fertilizer. The process is known as *fertigation*.

Deep percolation, where water moves below the root zone, can occur if a drip system is operated for too long or if the delivery rate is too high. Drip irrigation methods range from very high-tech and computerized to low-tech and labor-intensive. Lower water pressures are usually needed than for most other types of systems, with the exception of low energy center pivot systems and surface irrigation systems, and the system can be designed for uniformity throughout a field or for precise water delivery to individual plants in a landscape containing a mix of plant species. Although it is difficult to regulate pressure on steep slopes, pressure compensating emitters are available, so the field does not have to be level. High-tech solutions involve precisely calibrated emitters located along lines of tubing that extend from a computerized set of valves.

Sprinkler Irrigation

Crop sprinklers near Rio Vista, California.

A traveling sprinkler at Millets Farm Centre, Oxfordshire, United Kingdom.

In *sprinkler* or overhead irrigation, water is piped to one or more central locations within the field and distributed by overhead high-pressure sprinklers or guns. A system using sprinklers, sprays, or guns mounted overhead on permanently installed risers is often referred to as a *solid-set* irrigation system. Higher pressure sprinklers that rotate are called *rotors* and are driven by a ball drive, gear drive, or impact mechanism. Rotors can be designed to rotate in a full or partial circle. Guns are similar to rotors, except that they generally operate at very high pressures of 275 to 900 kPa (40 to 130 psi) and flows of 3 to 76 L/s (50 to 1200 US gal/min), usually with nozzle diameters in the range of 10 to 50 mm (0.5 to 1.9 in). Guns are used not only for irrigation, but also for industrial applications such as dust suppression and logging.

Sprinklers can also be mounted on moving platforms connected to the water source by a hose. Automatically moving wheeled systems known as *traveling sprinklers* may irrigate areas such as small farms, sports fields, parks, pastures, and cemeteries unattended. Most of these use a length of polyethylene tubing wound on a steel drum. As the tubing is wound on the drum powered by the irrigation water or a small gas engine, the sprinkler is pulled across the field. When the sprinkler arrives back at the reel the system shuts off. This type of system is known to most people as a "waterreel" traveling irrigation sprinkler and they are used extensively for dust suppression, irrigation, and land application of waste water.

Other travelers use a flat rubber hose that is dragged along behind while the sprinkler platform is pulled by a cable.

Center Pivot

A small center pivot system from beginning to end.

Center pivot irrigation is a form of sprinkler irrigation utilising several segments of pipe (usually galvanized steel or aluminium) joined together and supported by trusses, mounted on wheeled towers with sprinklers positioned along its length. The system moves in a circular pattern and is fed with water from the pivot point at the center of the arc. These systems are found and used in all parts of the world and allow irrigation of all types of terrain. Newer systems have drop sprinkler heads.

Rotator style pivot applicator sprinkler.

Center pivot with drop sprinklers.

As of 2017 most center pivot systems have drops hanging from a U-shaped pipe attached at the top of the pipe with sprinkler heads that are positioned a few feet (at most) above the crop, thus limiting evaporative losses. Drops can also be used with drag hoses or bubblers that deposit the water directly on the ground between crops. Crops are often planted in a circle to conform to the center pivot. This type of system is known as LEPA (Low Energy Precision Application). Originally, most center pivots were water-powered. These were replaced by hydraulic systems (*T-L Irrigation*) and electric-motor-driven systems (Reinke, Valley, Zimmatic). Many modern pivots feature GPS devices.

Wheel line irrigation system in Idaho, 2001.

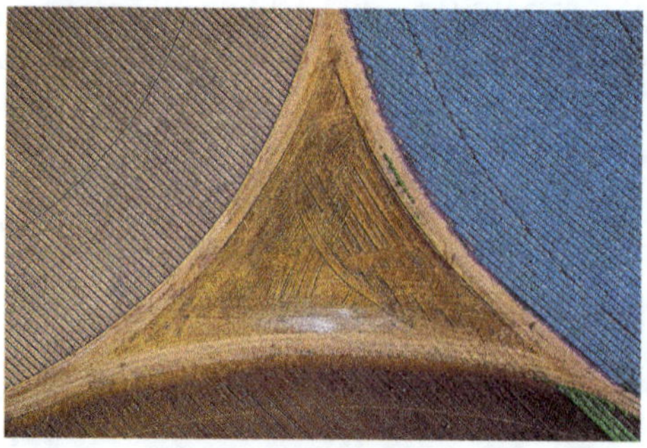

Center pivot irrigation.

Irrigation by Lateral Move (Side Roll, Wheel Line and Wheelmove)

A *series of pipes, each with a wheel* of about 1.5 m diameter permanently affixed to its midpoint, and sprinklers along its length, are coupled together. Water is supplied at one end using a large hose. After sufficient irrigation has been applied to one strip of the field, the hose is removed, the water drained from the system, and the assembly rolled either by hand or with a purpose-built mechanism, so that the sprinklers are moved to a different position across the field. The hose is reconnected. The process is repeated in a pattern until the whole field has been irrigated.

This system is less expensive to install than a center pivot, but much more labor-intensive to operate – it does not travel automatically across the field: it applies water in a stationary strip, must be drained, and then rolled to a new strip. Most systems use 100 or 130 mm (4 or 5 inch) diameter aluminum pipe. The pipe doubles both as water transport and as an axle for rotating all the wheels. A drive system (often found near the centre of the wheel line) rotates the clamped-together pipe sections as a single axle, rolling the whole wheel line. Manual adjustment of individual wheel positions may be necessary if the system becomes misaligned.

Wheel line systems are limited in the amount of water they can carry, and limited in the height of crops that can be irrigated. One useful feature of a lateral move system is that it consists of sections that can be easily disconnected, adapting to field shape as the line is moved. They are most often used for small, rectilinear, or oddly-shaped fields, hilly or mountainous regions, or in regions where labor is inexpensive.

Lawn Sprinkler Systems

A lawn sprinkler system is permanently installed, as opposed to a hose-end sprinkler, which is portable. Sprinkler systems are installed in residential lawns, in commercial landscapes, for churches and schools, in public parks and cemeteries, and on golf courses. Most of the components of these irrigation systems are hidden under ground, since aesthetics are important in a landscape. A typical lawn sprinkler system will consist of one or more zones, limited in size by the capacity of the water source. Each zone will cover a designated portion of the landscape. Sections of the landscape will usually be divided by microclimate, type of plant material, and type of irrigation equipment. A landscape irrigation system may also include zones containing drip irrigation, bubblers, or other types of equipment besides sprinklers.

Although manual systems are still used, most lawn sprinkler systems may be operated automatically using an irrigation controller, sometimes called a clock or timer. Most automatic systems employ electric solenoid valves. Each zone has one or more of these valves that are wired to the controller. When the controller sends power to the valve, the valve opens, allowing water to flow to the sprinklers in that zone.

There are two main types of sprinklers used in lawn irrigation, pop-up spray heads and rotors. Spray heads have a fixed spray pattern, while rotors have one or more streams that rotate. Spray heads are used to cover smaller areas, while rotors are used for larger areas. Golf course rotors are sometimes so large that a single sprinkler is combined with a valve and called a 'valve in head'. When used in a turf area, the sprinklers are installed with the top of the head flush with the ground surface. When the system is pressurized, the head will pop up out of the ground and water the desired area until the valve closes and shuts off that zone. Once there is no more pressure in the lateral line, the sprinkler head will retract back into the ground. In flower beds or shrub areas, sprinklers may be mounted on above ground risers or even taller pop-up sprinklers may be used and installed flush as in a lawn area.

An impact sprinkler watering a lawn, an example of a hose-end sprinkler.

Hose-End Sprinklers

There are many types of hose-end sprinklers. Many of them are smaller versions of larger agricultural and landscape sprinklers, sized to work with a typical garden hose. Some have a spiked base allowing them to be temporarily stuck in the ground, while others have a sled base designed to be dragged while attached to the hose.

Subirrigation

Subirrigation has been used for many years in field crops in areas with high water tables.

It is a method of artificially raising the water table to allow the soil to be moistened from below the plants' root zone. Often those systems are located on permanent grasslands in lowlands or river valleys and combined with drainage infrastructure. A system of pumping stations, canals, weirs and gates allows it to increase or decrease the water level in a network of ditches and thereby control the water table.

Subirrigation is also used in the commercial greenhouse production, usually for potted plants. Water is delivered from below, absorbed by upwards, and the excess collected for recycling. Typically, a solution of water and nutrients floods a container or flows through a trough for a short period of time, 10–20 minutes, and is then pumped back into a holding tank for reuse. Sub-irrigation in greenhouses requires fairly sophisticated, expensive equipment and management. Advantages are water and nutrient conservation, and labor savings through reduced system maintenance and automation. It is similar in principle and action to subsurface basin irrigation.

Another type of subirrigation is the self-watering container, also known as a sub-irrigated planter. This consists of a planter suspended over a reservoir with some type of wicking material such as a polyester rope. The water is drawn up the wick through capillary action. A similar technique is the wicking bed; this too uses capillary action.

Subsurface Textile Irrigation

Subsurface Textile Irrigation (SSTI) is a technology designed specifically for subirrigation in all soil textures from desert sands to heavy clays. A typical subsurface textile irrigation system has an impermeable base layer (usually polyethylene or polypropylene), a drip line running along that base, a layer of geotextile on top of the drip line and, finally, a narrow impermeable layer on top of the geotextile. Unlike standard drip irrigation, the spacing of emitters in the drip pipe is not critical as the geotextile moves the water along the fabric up to 2 m from the dripper. The impermeable layer effectively creates an artificial water table.

Water Sources

Irrigation is underway by pump-enabled extraction directly from the Gumti, seen in the background, in Comilla, Bangladesh.

Irrigation water can come from groundwater (extracted from springs or by using wells), from surface water (withdrawn from rivers, lakes or reservoirs) or from non-conventional sources like treated wastewater, desalinated water, drainage water, or fog collection. A special form of irrigation using surface water is spate irrigation, also called floodwater harvesting. In case of a flood (spate), water is diverted to normally dry river beds (wadis) using a network of dams, gates and channels and spread over large areas. The moisture stored in the soil will be used thereafter to grow crops. Spate irrigation areas are in particular located in semi-arid or arid, mountainous regions. While floodwater harvesting belongs to the accepted irrigation methods, rainwater harvesting is usually not considered as a form of irrigation. Rainwater harvesting is the collection of runoff water from roofs or unused land and the concentration of this.

Around 90% of wastewater produced globally remains untreated, causing widespread water pollution, especially in low-income countries. Increasingly, agriculture uses untreated wastewater as a source of irrigation water. Cities provide lucrative markets for fresh produce, so are attractive to farmers. However, because agriculture has to compete for increasingly scarce water resources with industry and municipal users, there is often no alternative for farmers but to use water polluted with urban waste, including sewage, directly to water their crops. Significant health hazards can result from using water loaded with pathogens in this way, especially if people eat raw vegetables that have been irrigated with the polluted water. The International Water Management Institute has worked in India, Pakistan, Vietnam, Ghana, Ethiopia, Mexico and other countries on various projects aimed at assessing and reducing risks of wastewater irrigation. They advocate a 'multiple-barrier' approach to wastewater use, where farmers are encouraged to adopt various risk-reducing behaviours. These include ceasing irrigation a few days before harvesting to allow pathogens to die off in the sunlight, applying water carefully so it does not contaminate leaves likely to be eaten raw, cleaning vegetables with disinfectant or allowing fecal sludge used in farming to dry before being used as a human manure. The World Health Organization has developed guidelines for safe water use.

There are numerous benefits of using recycled water for irrigation, including the low cost (when compared to other sources, particularly in an urban area), consistency of supply (regardless of season, climatic conditions and associated water restrictions), and general consistency of quality. Irrigation of recycled wastewater is also considered as a means for plant fertilization and particularly nutrient supplementation. This approach carries with it a risk of soil and water pollution through excessive wastewater application. Hence, a detailed understanding of soil water conditions is essential for effective utilization of wastewater for irrigation.

In countries where humid air sweeps through at night, water can be obtained by condensation onto cold surfaces. This is practiced in the vineyards at Lanzarote using stones to condense water. Fog collectors are also made of canvas or foil sheets. Using condensate from air conditioning units as a water source is also becoming more popular in large urban areas.

Efficiency

Modern irrigation methods are efficient enough to supply the entire field uniformly with water, so that each plant has the amount of water it needs, neither too much nor too little. Water use efficiency in the field can be determined as follows:

Field Water Efficiency (%) = (Water Transpired by Crop ÷ Water Applied to Field) × 100

Grapes in Petrolina, only made possible in this semi arid area by drip irrigation.

Until 1960s, the common perception was that water was an infinite resource. At that time, there were fewer than half the current number of people on the planet. People were not as wealthy as today, consumed fewer calories and ate less meat, so less water was needed to produce their food. They required a third of the volume of water we presently take from rivers. Today, the competition for water resources is much more intense. This is because there are now more than seven billion people on the planet, their consumption of water-thirsty meat and vegetables is rising, and there is increasing competition for water from industry, urbanisation and biofuel crops. To avoid a global water crisis, farmers will have to strive to increase productivity to meet growing demands for food, while industry and cities find ways to use water more efficiently.

Successful agriculture is dependent upon farmers having sufficient access to water. However, water scarcity is already a critical constraint to farming in many parts of the world. With regards to agriculture, the World Bank targets food production and water management as an increasingly global issue that is fostering a growing debate. Physical water scarcity is where there is not enough water to meet all demands, including that needed for ecosystems to function effectively. Arid regions frequently suffer from physical water scarcity. It also occurs where water seems abundant but where resources are over-committed. This can happen where there is overdevelopment of hydraulic infrastructure, usually for irrigation. Symptoms of physical water scarcity include environmental degradation and declining groundwater. Economic scarcity, meanwhile, is caused by a lack of investment in water or insufficient human capacity to satisfy the demand for water. Symptoms of economic water scarcity include a lack of infrastructure, with people often having to fetch water from rivers for domestic and agricultural uses. Some 2.8 billion people currently live in water-scarce areas.

Technical Challenges

Irrigation schemes involve solving numerous engineering and economic problems while minimizing negative environmental impact:

- Competition for surface water rights.
- Overdrafting (depletion) of underground aquifers. In the mid-20th century, the advent of diesel and electric motors led to systems that could pump groundwater out of major aquifers faster than drainage basins could refill them. This can lead to permanent loss of aquifer capacity, decreased water quality, ground subsidence, and other problems. The future of food production in such areas as the North China Plain, the Punjab, and the Great Plains of the US is threatened by this phenomenon.
- Ground subsidence (e.g. New Orleans, Louisiana).
- Underirrigation or irrigation giving only just enough water for the plant (e.g. in drip line irrigation) gives poor soil salinity control which leads to increased soil salinity with consequent buildup of toxic salts on soil surface in areas with high evaporation. This requires either leaching to remove these salts and a method of drainage to carry the salts away. When using drip lines, the leaching is best done regularly at certain intervals (with only a slight excess of water), so that the salt is flushed back under the plant's roots.
- Overirrigation because of poor distribution uniformity or management wastes water, chemicals, and may lead to water pollution.
- Deep drainage (from over-irrigation) may result in rising water tables which in some instances will lead to problems of irrigation salinity requiring watertable control by some form of subsurface land drainage.
- Irrigation with saline or high-sodium water may damage soil structure owing to the formation of alkaline soil.
- Clogging of filters: It is mostly algae that clog filters, drip installations and nozzles. UV and ultrasonic method can be used for algae control in irrigation systems.
- Assisting smallholders in sustainably and collectively managing irrigation technology and changes in technology.

CROP ROTATION

Crop rotation is one of the oldest and most effective cultural control strategies. It means that the planned order of specific crops are planted on the same field. It also means that the succeeding crop belongs to a different family than the previous one. The planned rotation may vary from 2 or 3 year or longer period.

There are very good reasons to rotate crops. Rotating crops usually means fewer problems with insects, parasitic nematodes, weeds, and diseases caused by plant pathogens. Rotations that include nonhost plants are effective for controlling insects like corn rootworm, nematodes like soybean cyst nematode, and diseases like root rot of field peas. When specific soil diseases are present, the length of time between growing the same or similar crop may vary from relatively short (one to two years for leaf blight of onions) to fairly long (seven years for clubroot of radish or turnip). Also, the rotation should contain some crops that are nonhosts or actually suppress the disease. Root growth may be adversely affected when continuously cropping to any single crop. This means that the crops may be less efficient in using soil nutrients and added fertilizers. In addition, rotations that include legumes may supply significant amounts of nitrogen to succeeding crops. A legume harvested for seed, such as soybeans, provides little N for the following crop. On the other hand, a multiyear legume sod such as alfalfa may well supply all the nitrogen needed by the following crop. Growing sod-type forage grasses, legumes, and grass-legume mixes as part of the rotation also increases soil organic matter. When you alternate two crops, such as corn and soybeans, you have a very simple rotation. More complex rotations require three or more crops and a five to ten-year (or more) cycle to complete.

Rotations are an important part of any sustainable agricultural system. Yields of crops grown in rotations are typically 10% higher than those of crops grown in monoculture in normal growing seasons, and as much as 25% higher in droughty growing seasons. When you grow a grain or vegetable crop following a forage legume, the extra supply of nitrogen certainly helps. However, yields of crops grown in rotation are often higher than those of crops grown in monoculture, even when both are supplied with plentiful amounts of nitrogen. Research in Iowa found that even using 240 pounds of N per acre when growing corn after corn, yields were not as good as corn grown following alfalfa with little or no N applied. In addition, following a nonlegume crop with another nonlegume produces higher yields than a monoculture using recommended fertilizer rates. For example, when you grow corn following grass hay, or cotton following corn, you get higher yields than when corn or cotton is grown year after year. This yield benefit from rotations is sometimes called a rotation effect. Another important benefit of rotations is that growing a variety of crops in a given year spreads out labor needs and reduces risk caused by unexpected climate or market conditions. Other benefits may occur when perennial forages (hay-type crops) are included in the rotation, including decreased soil erosion and nutrient loss.

ADVANTAGES OF CROP ROTATION

Crop rotation is the practice of growing a series of different types of crops in the same area in sequential seasons. Crop rotation gives various nutrients to the soil. A traditional element of crop rotation is the replenishment of nitrogen through the use of green manure in sequence with cereals and other crops. Crop rotation also mitigates the build-up of pathogens and pests that often occurs when one species is continuously cropped, and can also improve soil structure and fertility by alternating deep-rooted and shallow-rooted plants.

Crop rotation has increased in the south in the last 10 years due to the changing tides of the ever changing grain price. With the increase in corn acres across the south, as well as the increase in irrigation, we have seen a steady increase in yields. There are many studies showing yield increases of 10 to 15 percent in soybeans and corn when rotation is utilized.

Rotations also help with a reduction in nematodes, weeds and diseases. Northern Leaf Blight is a good example of a disease that has increased over the last several years, and can be reduced by rotating corn and soybeans.

Understanding the relationship between nitrogen (N) and crop rotation is very important when making N management decisions. There are several benefits to using crop rotation, including improved nutrient cycling, soil tilth, and soil physical properties; and enhanced weed control. Crop rotation also may influence the rate of N mineralization or the conversion of organic N to mineral N by modifying soil moisture, soil temperature, pH, plant residue, and tillage practices.

The incremental increase in N use over the past five decades, due to emphasis on maximizing yield, has led to a subsequent increase in N in the soil profile of some agricultural fields. Therefore, the influence of agricultural practices on water quality has prompted studies to develop best management practices to optimize the use of fertilizer N and reduce N loss to surface and groundwater. Crop rotation can play a major role in minimizing the potential risk of nitrate leaching to surface and groundwater by enhancing soil N availability, reducing the amount of N fertilizer applied, and minimizing the potential risk of N leaching.

Research on the impact of long-term crop rotation on soil N availability shows that planting alfalfa, corn, oat, and soybean significantly increased the mineralized net N in soil compared with planting continuous corn. Because soil N mineralization can effect yield, crop rotation thus can be used as a management system to enhance the soil nutrient pool, thereby reducing the fertilizer N input and minimizing the risk of leaching of excess N during wet weather.

A combination of conservation tillage practices and crop rotation has been shown to be very effective in improving soil physical properties. Long-term studies in the Midwest indicate that corn-soybean rotation improves yield potential of no-till compared with continuous corn. The reduction in yield of continuous corn in no-till is attributed to low soil temperature during seed germination, which is evident on poorly drained soils under no-till. Studies show that the poor performance of no-till corn following corn is more likely due to the previous crop than to surface residue conditions preventing early-season warming and drying of soils.

The use of a legume cover in crop rotation can provide a substantial amount of N to a succeeding crop. Research has indicated that seeding rates for legumes can be reduced by approximately one-third of that recommended for forage production when used as cover crops without sacrificing biomass or N accumulation. Also, the type of crop grown in the previous year can impact the efficiency of conservation tillage, especially for no-till systems, due to the kind and amount of crop residue from the previous crop.

References

- Principles-of-crop-science-environmental-sciences-essay, environmental-sciences, essays: ukessays.com, Retrieved 25 February, 2019
- Snyder, R. L.; Melo-Abreu, J. P. (2005). Frost protection: fundamentals, practice, and economics. Volume 1. Food and Agriculture Organization of the United Nations. ISBN 978-92-5-105328-7. ISSN 1684-8241
- Cropping-technique-2168-9881-1000e116: longdom.org, Retrieved 26 July, 2019
- Rodda, J. C.; Ubertini, Lucio, eds. (2004). The Basis of Civilization--water Science?. International Association of Hydrological Science. ISBN 9781901502572
- Topic-water-management, swmn, nafa: iaea.org, Retrieved 21 May, 2019
- Crawford, Harriet, ed. (2013). The Sumerian World. Routledge Worlds. Abingdon, Oxfordshire: Routledge. ISBN 9781136219115. Retrieved 2019-01-12
- Crop-Rotations, Text-Version, Building-Soils-for-Better-Crops-3rd-Edition, Books, Learning-Center: sare.org, Retrieved 13 August, 2019
- Advantages-of-crop-rotation: ncsu.edu, Retrieved 8 January, 2019

PERMISSIONS

All chapters in this book are published with permission under the Creative Commons Attribution Share Alike License or equivalent. Every chapter published in this book has been scrutinized by our experts. Their significance has been extensively debated. The topics covered herein carry significant information for a comprehensive understanding. They may even be implemented as practical applications or may be referred to as a beginning point for further studies.

We would like to thank the editorial team for lending their expertise to make the book truly unique. They have played a crucial role in the development of this book. Without their invaluable contributions this book wouldn't have been possible. They have made vital efforts to compile up to date information on the varied aspects of this subject to make this book a valuable addition to the collection of many professionals and students.

This book was conceptualized with the vision of imparting up-to-date and integrated information in this field. To ensure the same, a matchless editorial board was set up. Every individual on the board went through rigorous rounds of assessment to prove their worth. After which they invested a large part of their time researching and compiling the most relevant data for our readers.

The editorial board has been involved in producing this book since its inception. They have spent rigorous hours researching and exploring the diverse topics which have resulted in the successful publishing of this book. They have passed on their knowledge of decades through this book. To expedite this challenging task, the publisher supported the team at every step. A small team of assistant editors was also appointed to further simplify the editing procedure and attain best results for the readers.

Apart from the editorial board, the designing team has also invested a significant amount of their time in understanding the subject and creating the most relevant covers. They scrutinized every image to scout for the most suitable representation of the subject and create an appropriate cover for the book.

The publishing team has been an ardent support to the editorial, designing and production team. Their endless efforts to recruit the best for this project, has resulted in the accomplishment of this book. They are a veteran in the field of academics and their pool of knowledge is as vast as their experience in printing. Their expertise and guidance has proved useful at every step. Their uncompromising quality standards have made this book an exceptional effort. Their encouragement from time to time has been an inspiration for everyone.

The publisher and the editorial board hope that this book will prove to be a valuable piece of knowledge for students, practitioners and scholars across the globe.

INDEX

www.ingramcontent.com/pod-product-compliance
Lightning Source LLC
LaVergne TN
LVHW080621210926
848485LV00003B/193